Goat Song

My Island Angora Goat Farm

Susan Clark Basquin

J.N. Townsend Publishing
Exeter, New Hampshire

2000

Printed in Canada.

Published by

J. N. Townsend Publishing
12 Greenleaf Drive
Exeter, NH 03833
800-333-9883
www.jntownsendpublishing.com

(The chapter now titled "Navidad" appeared in a slightly different form in the *Wisconsin Academy Review*.)

Library of Congress Cataloging-in-Publication Data
Basquin, Susan.
Goat song / Susan Basquin.
p. cm.
1. Angora goat--Wisconsin--Washington Island. 2. Basquin,
Susan. I. Title

SF385 .B37 2000
636.3'985'0977563--dc21   00-036377

1-880158-28-0

For my nieces, Susan, Anne, and Jessie, and for my nephews, Peter, Lee, and Bill.

Acknowledgements

THIS BOOK is about a magical experience, a precious time in my life, that is necessarily the result of the many people, the kind and generous-spirited people, who make up the community of Washington Island. With their unfailing goodwill and supportiveness, these people made me feel—and they do still—as if we were all part of an extended family. There are many people—both islanders and non-islanders—to whom I owe so much and to whom I shall always be grateful. First of all, I want to thank my brother, Peter, and my sister-in-law, Jeannine, who owned the farm on the island and made the experience possible. And I want to thank Kit Basquin, my former sister-in-law, who has encouraged me in my writing from the very first. Cherie Weil, my friend from childhood, has been a constant source of support. Not only do I treasure her readiness to listen but also her perceptiveness, common sense, and extraordinary ability to see through to the essence of a situation. Toni Christenson and Gene Badeau provided constant support and encouragement and plunged in many times to lend a hand with the goats. Toni was the first to read my manuscript and offer suggestions, and Gene painstakingly converted the text from a word processor format into a computer program.

I could never have managed the care of the goats without the steadfast help, encouragement, and interest of John Herschberger, whom I shall never be able to thank adequately. My neighbors on the island, Barbara and Ray Hansen and their family, and particularly Dan Hansen, are people to whom I owe so much and with whom I have shared so much and still do. Donna Briesemeister and Ingrid Ryder, experienced sheep-raisers, were invaluable help whenever I needed them, which was often in the early days of my goat-raising. Kate Phillips and Ross Schilling and their daughters, Julia and Olivia, were eager participants in the goat experience, and Ross helped me countless times when the goats had puzzling medical problems. All I had to do was pick up the telephone. The veterinarians at Door County Veterinary Associates— formerly the Peninsula Veterinary Service—Joel Kitchens, Paul Haas, Joel Franks, G.R. Olson, and Randy Dietzel, never failed to respond graciously and patiently with advice over the telephone, even near midnight when I needed help with an illness or a difficult birthing. It would take too much space to list all the people to whom I feel so much gratitude for their help during my years on Washington Island. They indeed made the experience not only possible but a rich and rewarding one.

Last I want to thank Laurie Knowles and Dick McCord, who provided me with my first opportunity to write when they were publishers of the *Santa Fe Reporter*. Dick, as editor of the newspaper, was a superb teacher, and both have been caring friends for many years. Also, I want to thank Debbie Weissman and Fred Burrell, of Burrell and Weissman, CPAs, who encouraged me to work on my manuscript during slack times when I worked for them. Their interest and the supportive atmosphere they provided during a busy tax season were invaluable.

The Presence of Goats

IN THE HUSH of darkness dawn seeps into the sky, first barely visible, a pale orange line above the black pines that delineate the far end of the pasture. Outside, frigid air swirls and jabs, but when I step into the barn I sense warmth. It's not so much a few degrees of elevated temperature, but rather a warmth of spirit that mingles with the sweet peppery smell of alfalfa hay. I flick on the light to the does' side of the barn and see a dozen pairs of eyes regarding me benignly, calmly. Rachel is nestled in the bedding next to wary Gilda, her mother. Carmen sleeps through the interruption, her nose tucked low against her back legs. Ariadne stirs and yawns, then languidly pulls herself up to greet me. Musetta and her two-and-a-half-year-old daughter stand side by side. Fat Leonora already snuffles in the hay feeder, groping for remaining bits of fodder from last night's dinner. Octavian, son of Celeste, grandson of Tosca, hops onto the grain feeder near the door, the better to inspect me and any breakfast I may bring in, then jumps back down. Like his mother and grandmother, he jumps onto anything convenient; today he places his front hooves against my hip, and seems to search my face. I wonder if the trait was learned or inherited, as I wonder about many of the characteristics of the animals. The twins amble over, gazing at me with deerlike eyes. Like Octavian they're wethers, or castrated males, and are kept with the does because of their small size. Sprightly Kiri crowds in and Salome waddles over. Suddenly, silently, everyone is stirring, stretching to cast off sleep or getting up and moving toward me, nuzzling me, pulling gently at my jacket, eager to see if my next move will bring them a fat armload of hay. It's a trade, this morning ritual, and I come out the better for it. I give them breakfast and clean water; they lend me serenity and a good-natured perspective on the world. And if these should wear thin during the day, I have only to work with the goats for a while to recover a sense of well-being.

The care of Angora goats has shaped my life for a number of years, as has living on one of Wisconsin's Lake Michigan islands, reached only by water and air. Except for my dogs and cats and the goats, I live alone on a farm that is owned by my brother, who lives in Indiana, eight hours by car from the ferry dock, the jumping-off place to Washington Island. The goats are his; I am their caretaker and it is an experience that I treasure now and will look back on as immensely rich once my life changes. I have learned to work with livestock, mainly by doing it, accompanied by a generous amount of worrying and fretting, reading and asking questions, watching and worrying some more. With no veterinarian on the island, I've learned I can cope with medical emergencies and can stay up night after night birthing goats and still come back fresh and eager. I've learned that left to my own devices, I am never bored, but also that time can fill up with just the busyness of everyday living, even in this solitary existence, unless one harnesses it firmly. I've learned that like dogs and cats, goats have personalities. With a herd of ninety animals, more or less, at any given time, I focus mainly on the does. We get to know each other during birthing when I work closely with them; the goats draw me out of myself, amusing me, lending me their serenity, goading me into a response.

In the 1970s I read Jane Goodall's accounts of the varying personalities of the chimpanzees she observed. I admired her then for her determination to name what she saw and felt despite the scientific community's disdain. Now it's accepted that she was not only audacious and true to what she believed, but also that she was right. Later observations have shown the same: Cynthia Moss on kinship and friendship among African elephants; and Marcy Cottrell Houle, who observed what could only be called the grief of a peregrine falcon at the loss of his mate. Animals do show emotions, have feelings. Of this the goats have convinced me, particularly when they form friendships or cavort with companions or relatives after a long separation.

Animals may have feelings the way we do but theirs are clear-sighted, straight and clean; they haven't our capacity for fuzzy-headedness, our tendency for murking the waters. Live among them and they restore harmony in our world, and balance, and reflect to us what is important. They read our minds; sometimes I think the goats have eyes in the back of their head. Call it a sixth sense, but it's not too much different from a woman's intuition. And it's our intuition and our maternal instinct that pulls us, as women, into their world and melts the differences between species, if we'll let it. I gradually came to know the goats, to appreciate their intelligence, their quirkiness and humor, and I began to rely on the does at kidding time, depending on our teamwork as we launched their kids into life. I realized then that they and I were simply beings with different but equal roles in the mysterious scheme of life.

This life of mine is pared down, shorn of day-to-day rela-
tionships with people, populated instead with animals. The
dogs, cats, and goats intrude less than humans and give me
space and time for the landscape, the changing seasons, and
for myself. On this windswept tundra-like, winter's island, I
often crave color, hunger for it as if for food. Sometimes I find
it unexpectedly on a gray, bleak day when I notice the dry
grasses of unkempt fields suddenly, vibrantly the color of pale
sherry against the dull blur of leafless maples and birch and
the washed-out sky. When the winter landscape drains of life,
I turn inward, to mental pictures of other places, seeing in my
mind's eye the bright hues I miss. But most of the time the
landscape suffices, as does the sense of the unique existence
offered here. Still affected only little by the advertising and
mass consumption of our age, island life is sweeter than life in
other places and allows a generosity of spirit, a willingness to
help, that is foremost among the many solid, old-fashioned
characteristics of the islanders who form the small year-round
community.

One of the surprising joys of this life—this phase of my
life, because I do not know what is ahead—is its immediacy.
I'm plunged into it up to my ears, over my head. The life and
death of goats, births that go wrong, newborns that thrive and
grow, health, illness, all these shape my days, give substance,
meaning, pleasure and sadness and, not the least, intensity.
When I wrote for a newspaper, I wrote about other people's
pain, which I felt sometimes acutely and reported with a sense
of mission. But the experiences remained, in the end, vicari-
ous. I'd meet the deadline, later proof the galley, strip in cor-
rections, and paste the story onto the page. By the time the
piece was in print, I'd be researching the next assignment,
immersed in another glimpse of life secondhand. Now I live

in the midst of squalls that rise abruptly; I'm rained on and buffeted, pummeled by life rather than watching it from a safe distance. If I fail to act in whatever moment requires my attention there may be consequences. A kid trying to be born may become stuck in the birth canal; I may be able to help if I'm there. Sometimes the kid dies anyway; so far I have not lost a doe. If the billy crashes through the fence or breaks down the gate, I must fix it immediately if I don't want the females bred indiscriminately. Sometimes the one or two does I don't want mated because of their difficulty giving birth are impregnated anyway—it's a risk we take by not culling our flock, by not sending the less desirable animals to the sale barn, which is a euphemism for slaughterhouse. The consequences of our decision loom: long hours of birthing, usually near midnight, when I never know with certainty that I will succeed in pulling the baby, dead or alive, from the mother. Other times, if I don't respond when, say, at three a.m. I hear a faint cry from the direction of the barn, I may not find the kid who is trapped under a feeder or is caught in the fence. Not everything, of course, is a crisis, but with nearly one hundred animals, there is always at least one whose health is precarious, who bears watching. And there is the watching for the pure pleasure of it. My friend Toni is a psychotherapist and an organic farmer but defines herself, as well, as a dog watcher. The personalities of her five active animals provide a window into a canine world that fascinates her and which we can understand only partially. I am a goat watcher for the joy of it.

I am also a collector of perfect moments when the world seems to move into focus, each facet clear and in harmony with every other. As rare and precious as these instances are, the goat project has furnished more than one, although the single time I cherish most happened on a dark, midnight-black evening in winter. We had only twenty-one animals at the time and I had just fed them their oats and corn mixture. I walked outside through the low portal of their barn to the pasture to scoop any debris from their water container. A pure and silent snow blanketed the ground; large flakes fell gently, catching the light from the barn, which illuminated them against the darkness. Inside, an amber light mellowed the rough boards of the barn walls and softened the deep hay underfoot as well as the animals as they munched grain. All was still except for the sound of the goats snuffling and grinding corn. The hush of the moment with the clean snow falling quietly just outside the warmth of the barn conveyed a sense of peace and well-being akin to the manger scene in Bethlehem on Christmas night.

The Land

Late September the wild geese gather, tumbling high in the air, cavorting, dancing on currents, trying out formations, a marching band practicing. Soon they'll wing by, hundreds at a time journeying south, glimpsing no doubt the land they pass over. I like to imagine them peering down and seeing a loosely strung necklace of bristling green patches, each surrounded by a thin rim of white pebbly beach and floating on a lake of shifting moods. These are the islands that by an accident of glacial activity form a kind of bridge known as the Grand Traverse archipelago, which connects Wisconsin's Door County Peninsula with the Garden Peninsula of Upper Michigan. The Door Peninsula itself thrusts some ninety miles northeast from the city of Green Bay. Its edges are scalloped with fjordlike bays sheltered by limestone bluffs that rise on the western side of the peninsula as high as one hundred eighty feet above the waters of the bay, while the shore of the eastern, or Lake Michigan, side lies low and sandy. To geologists this protrusion into the lake is a classic example of a cuesta with an escarpment facing west; for me it remains my first breath-stopping example of the mystery and majesty of a wildness of spirit in land as well as water. I grew up in

Evanston, Illinois, on the shore of Lake Michigan, and was accustomed to houses set on careful lawns surrounded by hedges and sidewalks, to avenues running predictably north-south and streets east-west. In this ordered setting it was the surging lake, often stormy and apparently limitless, that thrilled me and unleashed my imagination. My family began spending summer vacations in Door County when I was still a child, and the forested bluffs edged by a white shore took their place beside the lake in that part of the mind that craves something free and unknowable. As if scooped out of the land to form the bays, chunks of pine-covered rock, most of them uninhabited, lie not far from shore and beyond the tip of the peninsula, forming the archipelago.

Of these islands, Washington Island is the best known. It lies six miles off the tip of the peninsula, across a strait of churning, icy water known as Death's Door. According to legend, it was first named by Potawatomi and Winnebago Indians who lost great numbers in battle during a storm. The name was translated to Portes des Morts by French explorers, who had their own reason to fear the channel: Among the many early losses was the first sailing ship on the Great Lakes, Robert de LaSalle's *Griffin*, which disappeared in 1679 crossing Death's Door laden with furs. It is thought that the ship had just set out from Detroit Harbor on Washington Island when it disappeared; its remains have never been found. Line squalls—the fast-moving storms that rise suddenly on the lakes and that come out of nowhere to whip the waters—currents that run counter to the wind; and rocky islands that create navigational hazards all proved too much for the lumbering ships, which could not turn quickly enough to meet the conditions posed by storms. In 1871 nearly one hundred vessels were wrecked or damaged crossing the Door; the next year eight wrecked in a single week. Less treacherous now because of modern nautical technology, the strait still presents a barrier, thus ensuring a certain peacefulness on the island compared with the peninsula, which draws millions of visitors every year.

In summer the peninsula's pristine villages set against green bluffs and sparkling water bustle with tourists. In winter, life turns inward. Summer residents abandon their large condominiums, which now stretch along the bluffs; visitors close up their vacation cottages and leave the resorts and campgrounds. Many of the restaurants and shops shut down for the season and the towns return to their year-round tally of 200 or 300 or 600 inhabitants.

Summer people come to the island, too, swelling its population, but relative to the peninsula it remains quiet and little affected by a surge of visitors. It is reached by car ferry, private boat, or small private plane, and the inconvenience tends to discourage many tourists despite a generous schedule of half-hour crossings by the ferry line's fleet. These drop off in the fall, paring down to only three a day in late December and a single daily roundtrip crossing in January, February, and March, when the strait freezes over and the single small ferryboat that can cut through the ice takes reservations in advance.

It's a watercolor landscape, this island, with colors, atmosphere, and contours defined by Lake Michigan, whose waves lap its shore and permeate its air. It's a landscape that J. M. W. Turner could have captured. Mist rises in spring and autumn, an effort to reconcile the disparity between the temperature of land and water; in winter putty-colored waves rear up, throwing white spray into the air, charging like stallions facing off.

There's an ephemeral quality to the colors, a transparency and translucence unknown in the Southwest, with its palette-knife sky and strong light and shadow. Here colors are paler, more delicate. Sunlight blends into shadow, lines are diffused, changeable; molecules of water dance more palpably in this air.

Roughly six miles on a side, the island is more or less a square patch of farmland and woods, marsh, and bluff, with natural harbors at three of its four corners. The island's first settlements grew up around these harbors in the 1800s. Washington Harbor, in the northwest quadrant of the island, is deep enough that it once provided a berth to large sailing ships and later, steamships. It became the first bustling commercial center on the island, with a warehouse, store, boarding house, and cabins on the harbor's west shore. All this vanished with the shifting of economic activity. Now, most of the year the harbor is silent except for the rhythmic crashing of waves during a storm; an overgrowth of cedar replaces the old commercial buildings and shields from view some of the vacation houses that have been built on higher land overlooking the water. Jackson Harbor, at the northeast corner of the island, remains the departure site for commercial fishing boats, which bring back whitefish and chubs. Small resorts with guest cottages, some of them many generations old, nestle there as well as at the small harbors at the island's southwest shore, while Detroit Harbor, more or less south-facing and home of the ferryboats, is the arrival gate for visitors.

Away from the wooded shore the island looks like a somnolent bit of Midwest farmland. Agriculture once flourished here in the form of potatoes and dairy and orchards but that era has closed, leaving as witness weathering barns and ancient cedar fence posts cocked at crazy angles like so many pieces of driftwood stuck in the sand. There are other remnants, too, clues to the layout of an agricultural economy: the rusted bits of barbed wire, sometimes whole strands, poking up from the ground or running through a tree that has grown around it, or anchored among the rocks of a decaying stone fence that snakes through a woods where it once marked a pasture. Now only a handful of farmers keep a few dozen head of cattle; a scattering of grassy hayfields are cut once or twice a season but seldom replanted with alfalfa or any other of the more nutritious crops used as animal fodder.

Resort cottages, summer homes, and the houses of couples who have retired to the island nestle along the perimeter, the wooded water's edge. The rest of the population lives for the most part in old frame farmhouses, spread scattershot among hay fields, stands of trees and former pastures, now grown over and wild. Main Street runs up the island from Detroit Harbor. Along it, strung out for a distance of somewhat more than a mile, are the school and community center, the island's single gas station, the grocery store, an all-purpose hardware store, bank, newspaper office, electric company, three taverns, and a couple of small restaurants that in winter may or may not open for a meal on a weekend.

The most salient feature of life on an island is its isolation, not only physical but also psychological in a way that affects its social development and, in turn, its history. An island fosters a different breed of personality, one that is relatively independent of the modern scramble for prosperity. Island culture sets its own values and priorities, which to outsiders may seem quirky. People here make do without a lot of money; except for a few trades, jobs are scarce during all but the summer months. Occasionally outsiders come to the island thinking they will open a business and make use of a pool of ready labor, underemployed and presumably eager for any wage. It usually doesn't work. A furniture factory opened some years ago and failed after not too much time. Off-islanders said it was because the workers would take off to go fishing on a nice day; islanders said the wages were too little. The truth, probably somewhere in between, has to do with priorities, as my brother learned firsthand. Peter started building a new hay and implement barn, planning to do most of the work himself with the help of a friend. The electric company sent its yellow vehicle equipped with a hydraulic posthole digger and bored twenty-eight deep holes for the supporting posts. My brother and Bill, his attorney friend, came up for a weekend of work, expecting that the people who had been contracted weeks before to pour cement around the posts would be there at the appointed time. The hour came and passed; no one showed up. Peter waited a reasonable time, then began telephoning. It turned out that there were two large weddings on the island that day and a soccer tournament on the mainland in which the island's seventh-graders were contenders. Not one contractor with a cement mixer was available. Furthermore, there was no crushed stone for making cement; the man who has the stone crusher also has a charter fishing boat and had been too busy taking out sport fishermen all summer to crush stone. What was important to the islanders was seeing their friends marry and their twelve-year-olds win the state soccer championship. My brother's postholes could wait.

When I first moved to the island, I had the sensation that I might fall off. I felt I was on a planet hurtling through space, on a thin piece of land adrift on water, and I, equally adrift, not rooted, had nothing to keep me from flying off the surface. Newly transplanted from New Mexico, where I had lived for sixteen years, I was not yet confident that the tendrils I put forth could find nourishment in the new soil.

In my mind's eye I see those first days as pale and drained of life. It was early August; the trees, heavy with summer, must have looked very green compared with the high desert landscape I had left. But the grass had dried yellow-brown as it often does in mid-summer, and I remember its lack of color and the pallor of the sky. Perhaps the comparison was not so much physical as mental. I knew the richness of New Mexico, its varied landscape, its history and culture. With its mix of Hispanic, Native American, and Anglo traditions it seemed as deeply savory as red chile sauce and as pungent as the aroma of pinon logs burning on a cold, clear night. What little I knew of the culture on the island, Scandinavian in its antecedents, seemed blond and bland and too pale and thin to hold me. Without wishing it, I was in danger of being spun off the earth for lack of finding a place where I could burrow in. In those first days I did not have the ability to appreciate or even recognize what was around me I had to learn gradually.

Josephine Herbst wrote in *The Starched Blue Sky of Spain* that she went to Madrid during the Spanish Civil War "because"; her reasons could not be put into conventionally rational terms. I came to the island in much the same way: because. My decision made little sense except to me. It certainly made no sense financially or professionally. One or two friends in Santa Fe thought it sounded like a tremendous adventure, but the others and those in New York and Chicago expressed their skepticism most eloquently by saying nothing. In my late forties I was leaving a fledgling but promising career in local journalism. My job at the newspaper in Santa Fe had ended with a change in management but my credits were good; I had written hundreds of stories and had won awards for some of them. I could have found another newspaper job in the state. But I wanted a new experience, something that would satisfy my restlessness for a time, something that might be the focus that I felt I lacked by not having a family or a deeply cultivated profession. And I wanted to regain a balance after ten years of a frantic pace working for a small weekly paper that consumed every waking hour. The island promised these things.

For several years my brother had encouraged me to leave my job in Santa Fe and live for a time on his farm on the island. It was not a working farm during that period but simply a vacation home for him and his family. Peter had remodeled the old frame farmhouse and covered it with cedar shingles. A low, gently curving stone wall led from the road to the house, which stood commanding a view of the 145-acre property that stretched across fields to the distant tree line on three sides. Beyond a sweep of lawn and set well back from the road, a large black barn had been remodeled to contain a guest apartment in the former horse stall and a studio in what had once been a chicken coop, although the central portion containing a hayloft remained unimproved. A smaller black building, a machine shed in former times, lay behind the large barn. Peter urged me to move to the island and use the studio for writing. The idea intrigued me only mildly. Then one day he mentioned raising goats. A longtime dream of my brother's was to re-establish the property as a working farm. He wanted to raise livestock to improve the soil, to plant orchards; he talked about exotic cash crops such as ginseng and English basket willows and commonplace crops to feed the animals. These plans were not new to me; for many years, over long dinners in Santa Fe restaurants, he had entertained me with his latest ideas for the farm. At one time he explained to me how musk oxen would be perfect for the farm because the young are born in May with a full coat of hair—ideally suited because spring is cold on the island. Furthermore, they prefer willow leaves, which fit conveniently with the basket willow idea, and they produce a fine hair that is harvested to make luxuriously soft fabric. I never doubted that my brother would carry out all his ideas, but other than amusing me over dinner, they had no bearing on my life. Next, it was Angora goats, which produce the fine fiber called mohair. They are easier to acquire than musk oxen, which live mainly in Alaska and Siberia, and are much less expensive. And at that time, mohair brought five dollars a pound. Peter had already planted alfalfa in one small field and bird's-foot trefoil, a forage goats like, in another field that could serve as a pasture. All he needed, it seemed, was someone on the spot to manage the operation.

It was at the point when the goats figured foremost in his plans that I was between jobs and wanted a change from newspaper work. What had been only mildly intriguing now seemed ripe with potential. To come to the island just to write would not have been enough, I recall thinking; to come to care for animals and write as well opened horizons I could hardly imagine. Peter's plan, however, was much more extensive than merely the raising of animals. Eventually we would process the mohair—wash, card, and spin it, using machinery as well as people hand-spinning it on spinning wheels. We would design mohair garments and have them woven. He'd set up a shop to sell mohair shawls, sweaters, and blankets as well as other island products. It would also be a workspace for weavers and spinners who would demonstrate their skill to tourists. He'd design a catalog of island wares and distribute it throughout the country. He'd be contributing significantly to the economy of a place he loved—giving something back to the island, he said, by creating jobs, work for women particularly during the long, quiet winter months. For me it would be a chance to develop a business with my brother, to work *with* someone rather than *for* someone. It would be an enterprise I would help create—I'd work at product development, marketing, management; it would be something I could truly shape and run. Ideally it would contribute to Peter's income and support me, perhaps for the rest of my life. And it would allow me, too, some time to write. Taken alone, the novelty of working with animals intrigued me sufficiently, but the larger plan offered real possibilities for a more secure future with some autonomy.

I accepted Peter's proposal: Come to the island, stay at the farm, and raise goats. I spent the next several months painting the interior of my condominium in Santa Fe to ready it for renters; I went to Guatemala for six weeks to brush up on Spanish—something I had long wanted to do but had never had time for. Then I took a temporary job selling advertising until I could find tenants to live in my house and thus finance my island adventure.

Landscape and Wind

THE LANDSCAPE and the wind are the two most pervasive features of the island. I had time to contemplate them that first year because Peter was not yet convinced of my commitment and wanted to wait before starting the goat project. I was disappointed. I was eager to start, and the decision made without any discussion rankled. But it was his money that would buy the goats and build their shelter, I rationalized, and he had a right to be prudent. So I set up a desk in the studio in the barn, made a fire in the woodstove on cold mornings and learned I could sustain myself mentally by writing. Mid-afternoons I would drive to the post office or the grocery store and back, then snap leashes on my dogs—Questa and Molly during those early years—and explore the country roads and woods. But however much I would escape in my mind to other places while I was writing, the wind and landscape drew me back again and again the way a favorite subject draws a painter. The landscape and the wind are constants of my life here, yet their kaleidoscopic changes furnish variety and, in the case of the landscape, often beauty so shimmering and ephemeral that all I can do is stop to take it in: the setting sun through a winter forest of slender birch glowing in the light; golden morning rays backlighting springtime trees; a slash of orange so vivid between horizon and dark gray clouds that I think I hear accompanying music, surely Vivaldi or Purcell. Throughout each day I gaze at dozens of variations on the same scene, hoping I can retain them forever in some recess of my mind. Will I be able to stop returning to these subjects once I get them down on paper?

The wind rose last night, awakening me, an entity that came out of nowhere. Suddenly it was simply there, roaring through the trees, howling around the house. When I was first on the island I once thought I heard the wind far away, although the air around me was still. The wind seemed to approach from a distance, ominously and relentlessly, but I could not be certain. In the next several minutes the wind did come and blew mightily all day long. Today's wind is from the southwest, less sharp than the northwest wind. It swirls and gusts around the house and barns, and batters the buildings from all directions. I close the various portals in the barn to reduce drafts, leaving only one entrance ajar in each pen, propped open just enough so that the goats can come and go. Mainly they stay where they are, hunkered down in the hay. The barn is situated so that they're protected from a north wind, and when it blows they often remain outside huddled close to the barn's south wall, peacefully unaware of the cold blasts absorbed by the building itself. The south wind, though, comes in through the low opening to the pasture. The goats find a spot inside, out of its path, and wait it out, which may take a day or two.

I'm always aware of the wind, its direction, its humming and howling, its intensity, and sometimes its absence. I rearrange my schedule for the wind. There's no point cleaning the barn because the manure would just blow out of the wheelbarrow before I dumped it. I won't finish trimming the boys' hooves because I don't want to shut the billy out in the wind. If I leave him in, he'll try to butt me and make a nuisance of himself while I work. I won't dig up the gladiola bulbs; any work in the garden leaves me with facefuls of dirt and scratched eyes. Besides, today's wind would merely scatter anything I pulled up and set aside. There's no point doing any task that would be more easily done once the wind abates. Inside, I start to clean the house. But the power fails and remains off; I can't use the vacuum cleaner and because the well pump is run by electricity, I can't scrub the floors until the electricity resumes. I can dust and sweep, though, and wait. Occasionally the lights flicker on but fail again. I build a fire in the woodstove to keep the house warm and glance at the battery-run clock. The clock has a wooden face cut into the shape of the island. It's a popular fixture here, where there are frequent power failures when the wind blows hard. We're used to it; it's a fact of island life. I know not to use the word processor during a windstorm or heavy rainstorm. The telephone also goes dead frequently and I know not to bother calling the telephone company from a neighbor's. Their line will be out, too, and service will resume in any event within a matter of hours.

Not able to complete my housecleaning, I decide to put up the new No Hunting signs Peter brought with him on his last trip. The Mercantile is almost always out of these signs and I half suspect it's the work of island deer hunters, who want to keep their hunting grounds from being posted. It must be frustrating to longtime islanders to see their favorite spots, on the wooded bluffs for example, carved up, bought, and built upon by summer people who immediately put up No Hunting signs. We're among those people, and I myself react with irrational emotion to deer hunting, although I know that it thins a deer population that might otherwise starve.

I pile the dogs into the van and drive to Peter's marsh property to put up a few more signs along the logging road that borders it. Then I drive along Michigan Road to inspect the signs I put up last year. Where some have been ripped down, I tack up new ones, fighting the wind the whole time. I circle down by the lake and am stopped by the grandeur of the water surging up, steel gray, spewing foam. It seems as if the lake has swollen, risen many feet, and on top of this, huge slow waves well up and crash, not so much against the shore as against each other. The conflict lends a chaos in slow motion, a measured, regular turbulence, a craziness that's somehow ordered. Today the lake seems to have its own rhythm apart from the noise and crash of the wind; it's a dance to music I can't hear. I drive home, check my friend Mary's dogs on the way because she is off the island for the day, then return with Lily, my large collie–blue heeler mix. I can't get enough of this drama just a few miles from the farm. On the horizon a tanker plods forward; I wonder how it fares in this weather, if the storm is as dramatic to its crew as it is to me. I'll return home to wait it out; I'll go to bed listening to it, hoping that it will subside so that I can return to my chores in the morning.

The landscape here is intimate compared with the boundless desert and distant buttes and mountains of the West. Most of the shore and the vast water beyond is hidden from view by woods. Everything else in sight lies no farther than the gray-brown blur of winter's trees at the horizon, often just across a forty-acre field. It's easy to think you have a handle on this small patch of farmland and forest, swamp and dunes and shore. But later you're surprised when you notice rocky seams across the land, bisecting deserted tracts of grassy weeds, popping up in the midst of woods, running along a road. The seams are the island's ancient stone fences that could be likened to the stitching in a patchwork quilt. Once functional, nowadays they no longer define pastures, fields, and orchards, or distinguish one farmer's land from another. Nearly melted back into the earth in places, they're vestiges of nineteenth-century Washington Island, witness and testament, irrefutable proof of a land at one time almost totally given over to farming, to orchards, to livestock and crops. At the same time, they hold an even more distant story, the tale of tropical Paleozoic Wisconsin, some four hundred million years ago near the equator. Honeycomb-like Favorites coral and Hayseeds coral that look like chains, branching Cladopora coral, and other fossils, cephalopods and snails, testify to warm-water reefs, once a part of this region's landscape. The fossils appear so abundantly in the rocks that compose the stone fences that they're commonplace to islanders and a temptation to visitors to carry away with them.

A foot or two under the surface the thin topsoil gives way to a deep bed of limestone known as the Niagara Escarpment, which stretches some nine hundred miles between eastern Wisconsin and Niagara Falls. Over time, the action of freezing and thawing causes this layer to throw off chunks that work to the surface to impede plant growth and plows. Farming began in a small way on the island in the late 1860s, and began to flourish when technology made possible the drilling of wells through the limestone layer. Relieved of having to carry water from the lake, the Norwegians, Danes and Icelanders who settled the island still had the backbreaking work of hacking rocks from the fields and piling them up at the edge of what would become orchards, cropland, and pastures for livestock. Even these sturdy Scandinavians must have wondered what they had taken on. But they persevered, carving fields, building fences from the stones wrenched from the earth. At one time stone fences also were built parallel to most of the island's roads, probably for want of a better place to deposit the rock. Sometimes these walls were dismantled and the stones crushed to build up and surface the straight roads that cross the island. Other fences were further demolished to build a dock to accommodate large barges that transported potatoes, a major island crop from the late nineteenth century to the late 1960s. A few stone fences still run the length of the roads in the wooded areas, but these are mostly decomposing and often hard to distinguish from the usual ridges and upheavals that characterize the terrain.

High stone walls line the north and south of my brother's farm and run along part of the western edge. Another bisects the goat pasture and is tumbling down from use over the years. At one time, barbed wire strung between cedar posts that were planted in the stone wall kept cows fenced in. Before we enlarged the original five-acre goat pasture to thirty, I spent days on the rock fence with barbed-wire pry and cutters jammed into my jeans pockets, thick work gloves on, dislodging bits of wire with the pry, then gnawing through the tough, rusted metal with the cutters. I'd walk the fence looking for a telltale strand, then pull what I found, bend it into a loop and leave it until I could return with a wagon to cart it off to the dump. But by no means did I extricate all the wire, nor could I. It poses a hazard to the goats. In full fleece they sometimes catch a piece that works into their hair and which I snip out. None of the animals, though, has ever been injured by the rusty wire because their dense mohair keeps it away from their skin. Nonetheless, I keep their tetanus shots current.

The goats love climbing on the rubble of these old fences and nibbling the chokecherry trees rooted in the stones. Withered chokecherry leaves are said to be toxic to goats, so before we fenced in the pasture I cleared the open field of chokecherry as well as juniper by pulling a mower for cutting brush behind an ancient red-and-gray Ford tractor. The chokecherries growing out of the stone fence were inaccessible, so I left those and hoped for the best. Predictably the goats nibbled, stripping the leaves before they withered and showed no ill effect. They continued nibbling tender branches until only sparsely branched dead trees remain.

One warm morning in late spring, as I returned home from walking Lily, I counted a dozen turkey vultures, wings partially unfolded, perched along the stone fence on the dead chokecherry trees and on old fence posts. I left Lily at home and returned to the pasture, binoculars in hand, for a better look. Naked red heads and necks, slightly hooked beaks, large black bodies with an almost six-foot wing span, they're sinister-looking at the very least. Spread down the fence line as they were, on dead trees and forgotten weathered posts, they probably enjoyed some form of avian conviviality. I moved closer, aware that if there were carrion nearby—a dead goat or a deer—they would be eating it rather than lounging above the rocks. It was humid that morning, and they were most likely cooling themselves in the breeze. As I approached they moved down the fence line but did not relinquish it, nor did I want them to.

Preparation

This morning the broad lawn is thickly yellow with dandelions, and nearly each flower has on it a honeybee, an Italian honeybee with gold and brown stripes. I'm happy to see the bees working the dandelions and don't mind at all these bright renegade blossoms on the lawn because I've always found them cheerful. Peter says dandelions are a sign of a particular deficiency in the soil, possibly copper; correct it, he maintains, and the weeds will vanish. But for now they are supplying pollen to the three hives of bees that I stationed one spring at the edge of the alfalfa field, just beyond the southwest corner of the lawn. I walk with peace of mind among the bees on their flowers; they are too intent on their work to pay me any notice, even if they could hear me, which I understand they cannot. Before I had bees—and before we had goats—the insects had a full range of wildflowers that grew profusely in the unkempt fields adjacent to the lawn. I'd look out over an expanse of yellow and orange hawkweed and Indian paintbrush for a week or two, then wild daisies, and later fields worthy of an Impressionist painting: purple spotted knapweed and Queen Anne's lace, as well as wild indigo, black-eyed susans, and purple aster. The landscape seems tame now; neatly sectioned into pasture, grass cropped closely, fence line mown. These days I like to look over the fields and remember their burgeoning wildness when we began preparing for the goats.

At some point during the first year Peter became sufficiently convinced that my interest in the goats was genuine, and we began preparations. Projects launch with expansiveness, with buoyant expectation and adrenaline surges. Nothing is too good; the sky's the limit. One wakes up eager for what the day will bring; details are mere trifles best dispatched quickly. I was excited about the goats, excited as I've seldom been in my life. Only matching it was the expectancy the summer before my first year of college. My role in launching the goat project was a new kind of adventure that brought novelty and challenge. In the past I worked for other people in well-established companies in which the role I played came to me ready made and carefully delineated. I've known men— worked for some even—who leaped from one challenge to the next, intoxicated by the new. They exuded energy and enthusiasm and often left a wake of unfinished business to be coped with by associates while they plunged into the next venture. Now I experienced firsthand the irresistible pull of new beginnings.

Peter ordered the goats through Susan Waterman, a breeder of registered stock just outside Madison, Wisonsin's state capital. We would not buy her registered animals but rather twenty commercial-quality does of a lot Susan was bringing up from Texas to promote Angora goat farming in the state.

My brother painted his plans in broad brush strokes, and in those days he let John fill in the details. Years earlier they had remodeled the farmhouse and barn, making a guest apartment out of the horse stall and a studio in the former chicken coop. Peter drew up the plans and John did the carpentry. Now they would transform the machine shed into a goat barn in much the same way. Most of July passed and I worried more every day that it would not be done before we picked up the animals at the end of the month. The machine shed looked hopeless. Built in the early 1900s at the time the house was constructed, its weathered boards hardly sheltered the stash of miscellaneous and historic farm implements and accessories kept there. One weekend Peter came up to the island and we attacked the shed, hauling away all manner of rusty vintage equipment. The dimly lighted, dusty shed had always been so jammed with discarded treasures and junk that I could not go into it without having to climb over objects or rearrange them. It was all a jumble: buckets full of bolts and screws, all thickly coated with rust; a turn-of-the-century machine for cutting hay that Peter later donated to the island's farm museum; an old car; a sailboat Peter bought secondhand and never had time to use; old pitchfork heads and handles, no longer joined; chunks of rusted equipment beyond my powers to identify; old brooms, tubs, cauldrons, a butter churn, several old metal and wood wheelbarrows and many other items in an assortment, all rusted, dusty, and grimy, and very likely priceless to the right person. Much we carted in several truckloads to the farm museum; the rest to the dump.

After a weekend of hauling, air thick with dust that had lain undisturbed for half a century, the shed was ready for John's attention. But first there was a pasture to be made. Stretching south beyond the lawn lay an untamed field of quack grass, spotted knapweed, timothy and, sparsely, the yellow flowers of the bird's-foot trefoil John had planted several years before. The field was full of generations of dried weeds and grasses, a pale, bleached-out thatch dotted at certain times of the year with wildflowers. Out of this thirty-acre expanse John marked off five acres and pounded metal fence posts into the rocky soil at eight-foot intervals. After research into the sparse information on fencing Angora goats and several calls to Susie Waterman, we determined the number of wires we needed, and their spacing. I ordered plastic insulators to put on the posts to hold the electrified fence wires away from the metal. These came in two parts, one to be screwed into the other, both stiff and hard to handle. So John, who invents whatever he needs, fashioned a wooden device with handle that fit over one part of the insulator, allowing me to screw the two parts together easily.

With bags of yellow insulators I set out one July afternoon to mark each post in seven places, then screw on the whole lot. It was tedious work, and slow; the day was hot and humid; the black flies buzzed and then bit. Toward evening as I worked my way around the five acres I began to feel unreasonably discouraged. I planned to finish the job that day and it wasn't happening. I could work by flashlight but realized the absurdity. When I wrote and edited at the newspaper, I stayed up all night if it was needed to finish a story or to paste up a section. All of us did. It seemed normal just to persevere. But here I was learning the difference between outdoor and indoor work. Here it was my physical endurance and strength that were challenged rather than mental stamina, and more difficult to comprehend, I did not have to finish the job in the time I allotted.

Next John and I strung the Polywire, a fencing material made of strands of plastic twisted with thin wire. Easy to use and to repair, it's cheap and good as a temporary fence. While John walked the fence with big spools of wire on a broomstick, I followed and guided each strand through an insulator.

Once we had strung up the fence, John connected it to the solar-powered fence charger he had installed a few years before around a patch on the property where he grew English basket willows.

I still despaired of the machine shed. Cleared of its debris, it was just a minimal shelter with cement floor. Its transformation looked impossible to do in the little time left. Adequate shelter is important for Angora goats in northern climates, because unlike sheep, they do not remain outside in the rain and snow and can get sick if thoroughly drenched, particularly in the six weeks or so after shearing. It was still July, but August was approaching with its cooler nights and sudden thunderstorms and soon enough, the time to shear. John quickly divided the main space into a goat pen and hay storage area. He installed insulation in the goat area and put up plywood, masking the weathered boards and preventing drafts. He cut a Dutch door between the pen and general storage and hay area, the walls of the latter left rough with great gaps between the old boards. But the goat pen was as snug as necessary for animals that need a certain amount of ventilation. Last, John made a hole in the roof for a cupola of his own design, to be roofed in copper. It was a device for added ventilation, which is important for goats and other animals that generate damp heat and can develop pneumonia.

Last, John fenced off a small area within the five acres. Our goats would have just arrived in Wisconsin from Texas where they lived on a dry, scrubby range. They would be unused to grassy pastures, although ours was beginning to dry and was decidedly less than lush. Until the goats adjusted, we would have to monitor their access to grass in order to prevent scouring, or diarrhea, or worse, bloating, which can be fatal in ruminants.

John is tall and quiet, and calm in the way of a person who is confident of his work. He knows he can figure out how to do almost anything in the way of carpentry or mechanics and has an instinct for farming. He's generous-spirited as well and sees goodness in others. John worked on the ferry line for years before starting his own contracting business, and his association with my brother has been long, dating to some of John's first major jobs. His is a hardworking, enterprising family. Patty, John's wife, helps with his business and has her own business painting designs on sweatshirts. His sister owns and runs Sievers School of Fiber Arts on the island, which draws students and teachers nationwide, and her husband runs the loom factory associated with the school. It was this connection that prompted John to grow English basket willows on Peter's property for use in the basket-making classes at the school and to fill mail orders. After work and on weekends, spring and fall, John spends hours on his knees thinning the clumps of willows or harvesting them.

John keeps only a few horses now, boarding them for other people, but at one time he had pigs, a cow, and chickens. He'd love to farm, he told me more than once, and was interested in the goat project from the first. He quickly became my greatest source of support, the person I turned to. But I did not know this when we first embarked on the venture; I was only glad that he wanted to come along with Peter and me to drive to Susie Waterman's, five hours away, to pick up our new flock.

The weather was clear, the air still and fresh that Friday morning. My sister-in-law, Jeannine, and my small niece, Anne, were at the farm along with their houseguests. We picked up John before seven and headed toward the ferry dock, the van pulling the two-horse trailer I borrowed from a local farmer. When I called around to arrange for something to transport the goats, I had little idea of what I needed. We would be returning with twenty yearling does, but just how tall they stood or how bulky they were was merely a guess. I had not yet actually seen an Angora goat at that point and I suspected my brother had not either. Would they fit in? I hoped so. We could stuff any extras into the back of the van or on my lap, we joked.

As we drove past Madison into rolling countryside I gazed at the lush green grass. The soil is deeper in that part of Wisconsin than it is on the island; pastures and lawns remain verdant even through mild droughts whereas on the island a few days without rain turns the grass yellow; weeks without rain, and the lawn browns and crunches underfoot. By mid-summer there's generally no need to mow a lawn except to trim the weeds that grow regardless of drought conditions. Compared with the pastures near Madison, our dry patch would pose no digestive problems to goats accustomed to the Texas range.

We found Susie and Clark Waterman's place, Odyssey Farm, tucked among gentle hills south of Madison. Susie, a slim, energetic woman with a doctorate in microbiology, had traded a career in science for raising goats. Her husband, Clark, made stained glass pieces and furniture but also lent a hand with the animals. They invited us to lunch around their kitchen table, and while we munched tuna salad sandwiches and potato chips, Susie summarized the basic care of goats. Hardly touching my lunch, I scribbled notes about wormers, delousers, feed, and other fundamentals. I'd read the one book that was available at the time on Angora goat raising in northern climates, *Raising Angora Goats the Northern Way* by Susan Black Drummond. I had pored over it and made notes. But without firsthand experience the names of diseases and medications, wormers and delousers were just so many new technical words that were hardly distinguishable. I wrote without comprehending, hoping I could decipher my scrawl later. I also made a mental note to call the nearest farm-animal veterinarians, the Peninsula Veterinary Service in Sturgeon Bay. They might have some useful information for me, and, in any event, it would be good to make myself known. It didn't take clairvoyance to predict I'd be a steady client.

After lunch Susie took us out to see the goats.

"Why don't you choose the twenty you want?" she suggested.

Choose the ones we want? I looked at my brother. He didn't register the least expression. Peter has the ability to become an expert instantly at almost anything. And if he isn't quite qualified yet, he never lets on. I am at the other extreme. I'd read and reread descriptions of what to look for in a goat. But like rules of grammar for a foreign language that slip one's mind at the first attempt to speak it, the traits I knew I should be attending to vanished from my memory. You're supposed to look at the legs, I remembered, and at the back. The face should have good hair covering but not too much. The horns should be just so. The fleece should be thick, lustrous, and heavy, and should grow down to the ankles.

Most of these qualities are the product of selective breeding started in Turkey. The forebears of Angora goats probably lived in the Himalayas, scampering up and down rocky terrain, which kept their hooves nicely worn. In that arid climate, which possibly kept internal parasites to a minimum, they ate scrubby vegetation, evolving to browse like deer rather than graze like sheep and cows. Smaller animals than today's Angora goats, they had long silky hair that was both warm and durable. It's easy to imagine women centuries ago keeping their goats as household pets, washing and combing their luxurious hair. The goats made their way to Persia, probably with caravans of traders; eventually they arrived in Turkey, where they flourished in the Ankara region, which gave them their name. The Turks spun yarn and wove fabric from mohair but did not allow export of the raw fleece until nearly the mid-1800s. When the Turks finally exported the raw material to Great Britain, among other countries, the English developed their own technology for spinning it, and the demand grew rapidly. To supply this market, Turkish goat keepers began breeding their goats with the larger, hardier Kurdish goat, breeding the offspring back to the pure Angoras for several generations in order to arrive at a larger animal that still produced fine hair.

Despite Turkey's guarding of its prized livestock, a few animals were taken to South Africa. According to one story, a reigning sultan agreed to sell some stock to South Africa but instead of shipping billies, he sent castrated males along with females that had not been bred. One doe, however, turned out to be pregnant and gave birth to a buck kid. Thus the South African mohair industry started.

In 1849, James Davis, an American from South Carolina, brought Angora goats to the United States. He had advised farmers in Turkey on their cotton crop, and the goats, five does and two billies, were a gift to show appreciation for his help. Other shipments of goats followed, but the fledgling industry in this country was soon disrupted and all but destroyed by the Civil War. By the end of the war, however, some animals had been taken to the Southwest and flourished, particularly in the southwestern part of Texas on the Edwards Plateau, where the climate is dry and the vegetation similar to the grasses and scrubby browse of Turkey. Large herds of several thousand became commonplace and spread to Oklahoma, New Mexico, and Arizona. By 1900 Texas was the second-largest producer of Angora goats after South Africa.

Angora goats were virtually unknown in the Midwest until 1979, when a Canadian company sold several thousand to Michigan farmers in an arrangement to supply the company with mohair and kids. The Canadian company failed, but by that time Angora goats were well established in Michigan in small herds of ten to several hundred animals. Drummond, an early goat farmer in Michigan, answered the need for information tailored to the upper Midwest climate by writing her book, first published in 1985. My dog-eared copy was literally never out of my sight the first year with the goats. It guided me through any number of crises, answered questions, scared me when I read about illnesses and possible mishaps—and soothed me when I needed reassurance. At least one other resource for Midwest goat raisers has been published: *The Angora Goat: Its History, Management and Diseases*, by Stephanie Mitcham Sexton and Allison Mitcham. And the classic book for anyone raising Angora goats remains Jean Ebeling's *The Angora Goat Book and Guard Dogs*, gleaned from her deep and long experience as an important goat rancher in Texas.

When I first saw our goats—or the flock of Texas goats that would furnish our small herd—they looked almost miniature. Standing on all four feet, their backs reached not much above my knee. Not having been shorn since January, the spring shearing time in Texas, they were covered with long, dirty ringlets. Angora goats have oil in their fleece, rather like lanolin in sheep. The greasiness of the fleece is partly an inherited quality but can also be produced by diet. The oil picks up dust and dirt, turning the outer hair a dingy gray, but part the locks and there's lustrous creamy white for most of the six inches or more of the staple. Angora goats grow their hair at a rate of an inch a month and are shorn twice a year, in the spring and fall. Let the fleece grow longer and it will begin to mat and pull out, and much of it will be wasted. Also, shearing helps rid the animal of the lice that are a persistent annoyance, even in the most scrupulously cared-for flocks.

Peter and I plunged in, picking out our goats, grabbing at horns, hoping we were choosing well. Susie helped us, gave advice and approval or told us when she didn't think a particular goat looked all that good. The task was relatively easy because the Odyssey Farm setup allowed groups of animals to be confined in a fairly small area. Susie's barns divide into many sections, most of them with their own small plot of pasture, which is an important feature of goat raising, particularly when there are numbers of goats of different sizes. Like any herd animal, goats fare better in groups but they impose a hierarchy based largely on size and age, but also on personality. They are tolerant of each other and quite amiable when there is plenty of space, but when crowded at the feeding trough, the smaller, less aggressive animals will lose out to the bigger ones. Our first barn was one big space; with only twenty yearling does of more or less similar size, it was adequate for the time being. Once kidding begins, however, there can never be enough separate areas. Ideally there is one for moms and young kids and a separate nursery area for mothers and younger kids of less than two weeks. There should also be a place for kidding pens where individual mothers and their kid or kids are confined for several days and another area for pregnant does just ready to give birth. And, of course, an entirely separate pen and pasture are needed for the billy. Our setup was never perfect; Susie's was the best I've seen. She also had a number of lightweight, portable panels for confining the goats temporarily, and she used these to keep our animals separate before we loaded them into the trailer.

"Do you want to see your billy?"

That was Susie's next question. Our billy? Peter and I had agreed not to get a registered billy until closer to the time to breed the animals; probably late October. We didn't have a pen for him, or a shelter. I glanced up at Peter but he had already turned to walk with Susie toward another part of the barn. He hadn't told me he'd changed his mind. There was nothing to do but trail along.

The billy was truly magnificent. He was the offspring of the original registered stock that Susie bought from a breeder of carefully selected, top-quality animals in Texas. His hair was thick and curly; his sturdy horns curved back and slightly outward. He was just a year old but was considerably larger than the does, and had a macho swagger.

In Angora goat circles, it's the practice to upgrade a herd by breeding commercial-quality does with registered billies. Registered animals are more expensive and should show better qualities through generations. Registration, however, is no guarantee that an animal will be superior, and likewise, many commercial-quality animals show all the desired characteristics one could wish.

Last, Susie gave me a demonstration of hoof trimming. Angora goats have two-toed hooves that look very much like the hooves of deer. Their tracks in mud and snow are similar. But the hooves must be trimmed regularly, about every three months; if not, they grow awry and can eventually cause lameness. Hoof trimming is also a good time to inspect for foot rot, a bacterial infection that spreads among flocks in damp climates and is difficult to eradicate.

Taking a goat by the horns, Susie deftly plunked the animal down on its rump, its back resting against Susie's legs. She bent over the doe, first scraping manure off the surface of the hoof with the point of her shears, then paring away a flap that had grown folded over one toe. Then she did the next toe, and the next hoof. Goats on the range in Texas need fewer hoof trimmings because the dry, rocky soil wears them down sufficiently. Usually these goats are trimmed only at shearing time. Susie had already trimmed some and was prepared to trim the rest, but I wanted a chance to practice. She handed me the shears. I knelt down and grabbed a hoof while she steadied the goat. The doe squirmed. Obviously she's aware I don't know what I'm doing, I thought. I'd be nervous, too. The hoof was hard and dry. And tough. The foot shears hardly worked for me. I proceeded slowly and carefully, carving off a little at a time.

"If you cooked the way you trim hooves, we'd all starve," my brother volunteered.

"Do you really want to go home with maimed goats?" I replied.

I worked on the last hoof and decided I'd just have to muddle through myself once back at the farm. I had the general idea. Goats can bleed to death from a bad cut during hoof trimming, I'd read, but with my conservative approach, it wasn't likely. Probably, though, I'd find myself trimming more often until I got the knack of it. I was relieved, too, that Susie was so friendly and interested in how the goats she sold fared. I'd be calling her often.

Late that afternoon, we pulled away from Odyssey Farm. We were silent compared to the morning's journey. I mentally reviewed the day's events and hoped the information on goat care was organized in my notes more clearly than in my mind. I reviewed the many words of caution: Don't leave loops of twine from hay bales around because goats can hang themselves; remove any handles from pails or buckets you leave in with the goats so they won't catch their horns; be careful of too much lush grass at first because the goats might bloat; check for runny noses, a sign of pneumonia; always check for scouring. It seemed a hazardous business, more than I realized. Any misstep, it seemed, could be disastrous, even deadly.

We stopped along the way to check the goats. I walked around to the back of the trailer hesitantly. Would I find goats carsick and prostrate? Would they be terrified, milling around, rolling their eyes, ready to stampede? I opened the door and looked in; they simply peered back at me, looking mildly interested. Above all, they seemed in some way cheerful. Their eyes registered inquisitiveness. Most were settled on the floor of the trailer; all gazed at me alertly, even brightly. At that point, they won me, completely and irrevocably. And I was more than reassured they'd make it safely to the farm.

Next, we stopped at Farm and Fleet, one of those huge stores covering whole city blocks that sells everything for farm, household, and garden, from tools to clothing to medicine. I perused the livestock section and found a new world, one of animal husbandry—of equipment and preparations for every aspect of animal care. I wasn't sure where to begin. Susie had mentioned Combiotic, a brand of penicillin, so I put a bottle in my shopping basket. I didn't see any of the wormers she mentioned. Peter picked up four shallow pans and some plastic scoops. These would serve for giving water and dishing out grain to the goats until we could find and install the proper receptacles.

We pulled up to the ferry dock in time for the nine o'clock boat. When we reached the farm, Jeannine and her friends were waiting to help unload. Peter backed the trailer to the barn door and we began leading out the goats, who seemed enormously self-possessed for having just traveled five hours to an unknown destination. We put down pans of water and left the goats for the night shut securely in the barn. As I glimpsed them a final time before going to bed, the billy stood with regal mien in the midst of his does, the lordly protector of the females who lounged contentedly at his feet.

Settling In

I awakened that first morning after the goats' arrival with the childhood Christmas-morning promise of happy surprises waiting. Dawn streaked the eastern sky with pink-orange. I dressed, throwing on jeans and a cotton shirt, slipping into socks and running shoes, and rushed downstairs. Coffee could wait. I hardly gave a thought to it, or to the dogs who, already excited by my exhilaration, whimpered to go out. I ran to the barn; Peter was not far behind me. We joked, gaily calling to each other. The adventure was beginning, shared with a person in whom I had utmost confidence. I would supply the day-to-day labor for the first five years and would draw no salary, but would have the money from any sale of fleece. Peter would finance the operation and guide it; I would learn from him while gaining my own experience in the business. At the end of the five-year period we would evaluate the venture and if there was a profit I would begin taking a salary. An intriguing future seemed contained in those bright little goats, and part of the beauty was the opportunity to develop new skills in association with my brother.

I opened the top of the Dutch door to the goat pen and looked in. Twenty-one pairs of eyes peered back at me, alert, mildly curious. The billy stood as he had the night before, playing the sultan to his harem gathered around him. I walked in, careful to move slowly without abrupt gestures. Seeing the goats shy away, I told myself not to expect miracles. It would be take time for them to become tame. Except for the billy the goats were Texas range animals, unused to human attention.

"Shall we do it?" Peter looked at me. Humor and enthusiasm flickered across his face.

"Let's do," I answered. And we opened the door to their pasture.

I held my breath—Peter may have, too—and waited. One by one the goats tentatively ventured out, taking a few stiff-legged steps and then a little scurry for several feet. They watched us; then, suddenly excited by nothing I could see, darted away. Soon a few began cropping grass. Finally they all milled around outside in a loosely knit group. We watched them and I marveled at having our own farm animals. They felt like mine, although they were Peter and Jeannine's. I would simply care for and get to know them. Just as I was reflecting on the goats, I noticed a blotch of fleece roaming outside the electric fence. Could it be? I looked again.

"Peter! Do you see what I see? Whoops, there's another."

"I'll unhook the fence."

Peter dashed through the barn and across the field to the willow patch, where the goat fence connected to the solar-powered fence charger. I watched as another goat and then another slipped through the electrified wires. As soon as the current was cut off, I climbed through and began chasing goats, who now were free to run throughout the whole five acres and beyond if they cared.

My brother is not an athlete. Chasing anything the first thing in the morning is not something he normally does. I jog—or in those days, ran—every morning, so I was a bit more up to the task. We began with the goats still in the smaller enclosure, running after those near at hand, catching them and carrying them back to the barn where, panting, we secured the door after them. Were the goats frantic or playful? I couldn't tell. They were surprisingly light, though perhaps the demands of the moment made their sixty or so pounds seem less. Jeannine came out with four-year-old Anne, and seeing that we needed help went back to get their houseguests. Soon we were all running in pursuit of errant goats. With two or three of us working together we could corner and catch many of them. Those outside the enclosure we herded to the point where they slipped back through the wires and scampered toward the narrow passageway into the barn. When all were safely contained inside, Peter announced it was time for breakfast. We'd go to the Sunset Resort and figure out what to do next over Icelandic pancakes.

What to do next was to call Donna, who has sheep. It was the first of many times I would call on Donna for help. Yes, she had a spare fence charger, the kind you plug in and run on ordinary current. Peter and their guest CJ rigged it up to the fence. The voltmeter showed eight kilovolts, better by far than the four kilovolts from the solar-powered charger. Perhaps between the willow patch and the goat enclosure there was just too much fence for the solar device, or perhaps the collecting panels were too old and worn. Again we let the goats out and held our breath. This time they stayed in. At least for the moment.

As much as I tried to push it away, doubt sidled in to mar the euphoria of that first day. I had begun to sense a gap yawning between my brother and me with respect to our roles in the project. Perhaps my expectation of equality in the venture was extravagant and unjustified; after all, he was putting up the money and furnishing the pastures and barn. In my fantasies I had envisioned a joint enterprise in which we would make decisions together, and it did not seem to be unfolding that way. I reasoned that perhaps I could regain what felt like a crumbling footing by solemnizing our relationship with money in a formal business arrangement. I had a little capital; I could invest it in the project and thus, I hoped, gain status. I suggested my idea to Peter. His answer was immediate and cool. "You come up with a business plan, and I'll consider it." I pursued it no more.

In retrospect I realize that my inability to move further with my idea was thwarted by our cultural differences. Having come of age in the Midwest in the late '50s, I was steeped in the female expectation of inadequacy, at best, in business and other traditionally male areas. And like many of us I gave my male counterparts that very message by my example. So there I was, hoping to be coddled and encouraged—and clearly not showing I deserved an equal place. And my brother, entrepreneur that he is, was thrown into business mode by my request and countered as he would in the marketplace, as if dealing with a potential adversary. If I had been a child of the '70s, I would scarcely have been discouraged by him. Rather, I would have accepted his response as a challenge to gain what I wanted. And in doing so, I would have merited that elusive equality I so much desired.

Later that day Peter, his family, and their guests left. As they drove away, I felt stranded on quicksand, a sinking feeling mostly in my stomach. I was caretaker of twenty-one living creatures who might or might not be successfully contained in their pasture. If they got out, would I be successful in getting them in again? And what else could go wrong? It was not that Peter or Jeannine knew any more than I about the goats, but there was security in numbers. Responsibility spread out weighs less heavily. Now the immediate responsibility was all mine.

The goats and I might well have been from different planets, we had so little experience of each other. They had been around humans more than I had been with goats, but the tally for both of us was only a matter of days. Furthermore, their interactions must have been frightening: shearing, inoculations, herding into a truck to make the long journey from Texas to Wisconsin. They were wary of me; I less so of them. They were small, first of all, a manageable size although they would grow. And they didn't bite. Goats have sharp upper and lower molars but in the front of their mouths they have only lower teeth. The upper part of the mouth is a hard gum or rim, which won't break the skin if the animal should decide to nip, something that happens seldom. The young billy, of a formidable size compared to the girls, was accustomed to people and was as gentle as a kitten. He came up readily and liked his nose rubbed and the base of his horns scratched.

Unlike many dairy goats, Angora goats are placid creatures that won't butt people. There's no need to worry about turning your back on a doe or a wether. The billies, though, are different, as I learned from experience. They're deceptively mild until they reach the age of two. At that time an urge for dominance takes over and they become aggressive, particularly during breeding season. It's not that they're bad-tempered or mean; they just like to challenge their caretaker. It was only at the second breeding season that our billy began rearing up on his hind legs, his head lowered and slightly to the side, and dancing at me as if challenging me to spar. If I came in with a bale of hay, he'd try to butt it, or if I brought a ladder into his pen to fix something, he'd work at it until I had to shut him away if I didn't want the ladder knocked out from under me. I tried to see his side of it; I told him I understood his predicament. The does ovulate during a twelve-hour period that occurs only every nineteen to twenty-one days, approximately from mid-August through January, and so the billy is under a certain amount of pressure to mate with them while he has the opportunity. For much of the breeding season, I keep the billy goats separated from the females so that we won't have kids too early in the year. Thus I'm a deterrent to satisfying his lust.

I tried all sorts of things to cure him of his threats: I shouted at him; I acted like a crazy person to get his attention, dancing around and hollering. I tried reasoning with him, explaining my needs and how I knew they differed from his. For a time a water pistol helped. I'd squirt him in the face with plain water when he threatened to butt. I felt like a Wild West cowboy, going about my chores with a pistol in one hand aimed at the billy.

An acquaintance who once worked on cattle ranches said that the problem was that the billy, like bulls on a ranch, perceived me as the dominant being and it was simply in his nature as a male animal to challenge that dominance until he won. It seemed true enough; the billy hardly ever challenged anyone else. John could work in the billy's pen without that pawing of the ground that signals a charge. I tried using one-and-one-half-inch PVC pipe; I'd bop the billy hard over his horns. He shook his head and looked stunned the first time, but later I broke many pipes working with billy goats and failed to make a lasting impression. Finally, I settled on an electric cattle prod.

Much later another goat raiser advised against making friends with billy kids lest they grow up knowing there is nothing to fear from their caretaker. Healthy fear of humans is a good thing in a powerful billy with an impressive rack of horns. But I did not know that in the early days. I felt strongly that if I treated the goats well, they would respond in kind.

My best use of time, I figured, would be spent making friends with the goats. After all, their hooves had to be trimmed every three months or so, they had to be given wormer for internal parasites regularly, and had to be treated for lice several times after the twice-yearly shearing. The more tame they were the easier it would be to herd them into an enclosure and catch them for whatever procedure was necessary. The less frightening I was to the goats, the easier my life would be, I reasoned. But I knew that it was all partly an excuse. I was intrigued by their hesitant inquisitiveness, their bright, quick movements, the calm that seemed to spread and deepen in the pen at night when they settled in the hay. And they looked so soft and cuddly. I wondered if I could get them to eat from my hand. I wanted to feel their soft noses nuzzling for grain. I imagined plunging my fingers into their thick, curling fleece.

I grabbed an empty red, green, and white Medaglio d'Oro espresso can, filled it with oats and corn, and went out into the small pasture. It was warm and sunny; birds twittered in the trees at the edge of the pasture; puffs of clouds nudged into the silken blue sky. It was perfect weather for sitting in the grass and getting to know goats. I crouched close to the ground and waited, hardly moving. They were wary and maintained a distance. Even the billy kept away. I moved slightly to settle myself into a sitting position and startled the goats, who scattered a few yards farther into the pasture. I kept still and slowly they edged closer, a few of the more curious ones taking the first demure steps. I raised my hand to hold out the coffee can of grain, but that slight movement sent them back. At feeding time in the evening they had not shown fear but rather crowded in despite my presence, greed overcoming any reluctance. I slowly scooped a handful of grain from the can and extended my arm. They backed off at first and stood watching. Then slowly they advanced, a couple at a time. Finally one doe reached me. With a preliminary dart back and forth and then another, apparently making up her mind, she decided to chance a closer look at my hand. In moments her soft nose pushed into my palm and she gobbled. I expected her to scamper away at any moment, but she ate until she finished the corn. This piqued the curiosity of the others, and the crunching of grain spurred their hunger until a few followed. Equally tentative at first, four or five of the bolder animals sashayed up to me until they, as well, delicately nibbled at corn and oats.

It was all carried out in silence except for an occasional awkward "Here, goat," as if calling a dog. I tried "Goatie, goatie," but that didn't suit. Then "girls and boys"; it rang somewhat better. I simply didn't know how to talk to goats; I felt self-conscious and was glad no one was there to hear me. To compound the problem, the goats all looked alike: two eyes and a long nose, two horns more or less curving back over masses of curls. Fortunately each goat was identified by a number on a round plastic disk in one ear and could be further distinguished, if one could read the language, by ear notches.

Most commercial goats are notched for identification

because ear tags sometimes fall out and the numbers on them eventually wear off. All registered animals, both here and in Texas, are not only notched but also tattooed inside the ear flap. The notches conform to a system whereby the numbers depend on the location of the notch and whether it is on the left ear or right—a bit like reading braille.

During this period Peter came to the island at least once and we shared a pleasant hour sitting in the grass feeding the goats. He loves animals and is good with them, and I was eager to introduce him to the delight of making friends with our Angoras.

I began noticing the numbers of the goats closest to me. Several sessions in the pasture with my red, green, and white coffee can drew the same goats to me. No. 120 was always there, friendly and rather sweet with a certain unassuming constancy. I called her Mimi. Another inquisitive doe was easily identified because her horns rose straight from her head at a slight outward angle as if defining a Spanish comb. I called her Carmen. Tosca, No. 107, was the greedy one, eager and funny and demanding. Soon she learned to climb against me, her hooves on my hip in order to get closer to the coffee can. Within not much time a few of the animals regularly pushed up to greet me, crowding and shoving for the grain can. Mimi even began following me in the barnyard like a puppy.

If there was pleasure that first week, there was also worry. The goats seemed to sneeze a lot. At least I thought it was a sneeze. They made little puffing sounds, especially at mealtime. Were they all coming down with pneumonia? Perhaps the change from a dry climate to a more moist one with cold nights was making them sick. Perhaps the stress of the trip affected them. I decided to wait and see.

In the late afternoon, as I was leaning over the door to

their pen and wondering about the sneezing noises, I noticed blood at the base of the horn of one of the does. Alone with twenty-one rather alien creatures, I often didn't know whether to worry, which may be the same as worrying. Feeling a bit silly because I suspected I was reacting with undue concern, I called Ingrid, who kept sheep and a dairy goat or two. Goats nick each other from time to time, she reassured me. But to prevent flies from laying eggs in the wound, I should dab some hydrogen peroxide on it. I caught the doe with some effort—she had not yet learned to trust me—and pressed a saturated cotton ball to the wound.

Outside, thunder cracked twice, then rain started pelting the barn roof. By that time the goats were penned up inside for the night. But the roof still had a hole cut in it for the cupola John planned to install. Sneezing—at least making noises like sneezing—and possibly coming down with pneumonia, my goats were now going to be drenched inside their barn. I had to cover the hole; it was the only solution. But what to do? I looked in the main barn, not sure what I was searching for. Then I glanced around the pump house and found a sheet of heavy clear plastic folded into a large square. It would do, at least if I could determine how to put it up. I studied the inside of the goat barn while the animals milled around me. Nothing looked easy there. The only solution was to get a ladder, climb up on the roof, and nail the plastic over the hole. Rain now gushed down, lightning flashed, thunder boomed in the distance. I'm not comfortable with heights, and the rather low goat barn seemed higher once I was on the roof. The wind picked up and my concern was now to secure the plastic so that gusts would not pry it away. I pounded in nail after nail and wondered if I harmed the roof doing so. But I was assured that at least I was doing my best for the goats.

Once finished, I felt amused at the twists this farming venture would take. The animals were now protected from a soaking and perhaps the dread pneumonia their puffing suggested. And I had just experienced climbing around on a rooftop in a thunder-and-lightning storm.

Pneumonia never developed. In not too much time, I discovered that the puffing was never a sneeze in the first place but was merely a goat sound, a kind of snort that's very similar to the snort a deer makes when threatened. And because goats are by nature somewhat aggressive with each other, they snort a lot.

The goats should be wormed right away, Susie Waterman advised over the telephone. I decided to carry out that operation at the same time I trimmed hooves. I'd heard dire things about wormers, some so strong a slight overdose can make an animal dangerously ill. The image of finding all the goats dead one morning appalled me, and that grisly picture flickered through my mind from time to time in those early days. I chose a deworming paste with the reassuring name of "Safeguard" and determined an appropriate dose based on estimated weight of the goats.

Concerned that I might overdose the goats—or not give them enough wormer to rid them of insidious internal bugs— I reasoned with myself about this and about the hoof trimming. After all, farmers have been worming livestock for generations. Goat ranchers have trimmed hooves ever since goats were removed from their native dry, rocky habitat. Forget that the book warned that a goat could bleed to death from a careless hoof trim. It was unlikely I would be anything but cautious as I wielded the hoof shears. For this initial challenge, however, I had not only moral support but hands-on help as well. Peter's secretary and good friend, Marcia, her husband, Bill, and their daughter, Jennifer, a pre-vet student, were staying at the farm for their summer vacation. Jennifer and Bill were happy to help me with the double operation, although Jennifer confessed she was more interested in small-animal veterinary medicine than farm animal care.

I studied the logistics. It seemed best to pen the animals in their barn, set up shop just outside the door to the pasture, and take the goats one by one. Once I'd trimmed them and wormed them, I'd release them into the pasture but not before giving them a handful of grain. I felt it was important to make the experience a positive one. After all, I was working toward easy management and happy goats; if I succeeded I might have goats scrambling for a hoof trim and a bit of deworming paste.

The first goats were easy. Friendly Mimi was the guinea pig. Still small enough to pull down easily into a half-sitting position, she rested against Bill, who steadied her while I worked slowly and carefully on her hooves. They were such a mess I hardly knew where to start. And they were hard. It had been several days since the rainstorm and the ground was thoroughly dry; the hooves had consequently dried to a daunting toughness. When I finished with Mimi's hooves, I gave her three cc's of runny paste that looked like Elmer's Glue, squirted into her mouth from a syringe. Then I offered her oats and corn. Bill and Jennifer were troupers. The trimming took hours, and for the most of it the three of us knelt or sat on the ground, which was already littered with shiny black pellets—goat droppings—which resemble the scat of deer. If either Bill or Jennifer had known the job could be done faster or felt they could do it more efficiently, they never let on. The goats, once down and held firmly by two people, gave little problem. It was only catching them that became an athletic contest after the few friendly does had been released into the pasture. In any group of goats there are always two or three who refuse to make friends or even cooperate however minimally. Brunehilde was one; she was aggressive and belligerent with the others from the first and not the least disposed to warm up to me. There were a couple of others, too, who were more timid than bellicose. Chasing them around the pen was frustrating. Worse, I could feel their anxiety escalate into terror; I worried one might break a leg in the melee. And it seemed much too hot and humid in the barn to run down our terrified animals. It was almost a year later when I attended a conference and show that I learned that shepherd's crooks are still an indispensable part of goat and sheep raising. I also learned to drop everything during a good rain, which softens the hooves, and trim. In those early days it took me all day to do twenty-one goats single-handedly. Now, after several years, I can trim one hundred goats—four hundred hooves—easily in a day and still have time for other chores. Not the least of it is that the goats know me; only a few do I have to snag with my shepherd's crook.

Patterns

MY LIFE HERE with the goats has become a design: the backdrop is the countryside; the parameters, a response to the seasons; the pattern in all its intricacy tells of my life with the goats. The pattern is a strong one now and much to my liking, partly because I've loved the experience that forged it. But it's had to be honed over time, the myriad details arranged and rearranged and sloughed off if unimportant, the true ones deepened by repetition. In those first days every action was a tentative element in the pattern and potentially carried equal weight with every other element. I did not know what would or would not serve and hardly had a sense of what was truly critical; but soon a pleasant rhythm established itself around the goats. Still entranced by their novelty, I would rush out to see them in the morning just as the sun spilled its light over the pine trees to the east. I'd skim through the dew-wet grass and feel exhilarated by the sweet, fresh smell of the air and the world that was all green and blue and golden. The goats were amusing to watch and undemanding to keep; a few pans of water during the day, a bale of hay and their grain at night, and they were content. They munched pasture, but it was never lush and was quickly eaten to a stubble. By the second week, I released them from the smaller enclosure into the entire five acres and I let them have their freedom at night rather than keeping them penned in the barn. There are no wolves on the island and the only coyotes exist in rumors circulating at the taverns. I soon learned that there's no worry that a solitary goat will wander off to explore; it simply isn't their nature. The little band moved in a loosely knit clump, slowly circling the pasture during the course of the day.

I began to be less concerned about the stray sneeze, occasional limp or runny nose that miraculously disappeared the next day. Every morning I'd inspect the goats for signs of scouring, which could warn of intestinal parasites or other diseases or simply a bit of indigestion from overeating. I'd check for runny eyes and noses and anything else amiss. I'd do it all very consciously, carefully ticking off the goats in my mind. Now, with experience, I'm hardly aware of the mental process as I automatically note the condition of individual animals whenever I come into their pen. I do this at least twice a day, at feeding times, but more frequently during extremely cold or extremely hot weather, and during storms. If a goat looks off, I make her move a few steps so that I can note any further sign such as a limp or sluggishness. I may take her temperature, and I note whether she's chewing her cud. There's little that's more satisfying than seeing all the does solemnly chewing of a morning. In the first days my inspections were very deliberate. If anything, I erred on the side of being overly cautious.

"One of the goats won't put her foot down, Ingrid, and I can't figure out why. I looked at her hoof and there doesn't seem to be anything wrong, like a stone between her toes. Do you have any suggestions?"

Fortunately, it was not the kind of call I had to make often.

Ingrid was very kind. If she was amused by my oversolicitousness, she never let on.

"I'd suggest you just keep watching her. Goats can be very sensitive. She may have been butted and is a little bruised," she said. "But it's good to check these things out," she added. "They always say a good shepherd attends to any problems immediately."

By the next day these little worries generally disappeared without any intervention. What did need my attention, though, was the fence. The fence charger was a duplicate of the one we initially borrowed from Donna and had a flashing light and a loud, regular click that I could hear faintly from the house on a very still night. When I entered the barn, I'd glance at it. If the light was not flashing and the click sounded different, something was wrong. The first time I noticed a malfunction was about ten-thirty one night when I went out for a final check on the goats. I loved seeing them settled peacefully in their barn. If the night was warm, I'd look for them outside by sweeping the pasture with the beam of my flashlight until I saw a cluster hunkered down in a depression, pairs of eyes shining at me. That night the goats were inside. Nothing amiss there, I thought. But the fence charger wasn't flashing. Probably there was no current circulating through the wires, although the charger was plugged in properly. I was concerned that a stray dog or one of the rumored coyotes might burst through the fence lines and attack a goat. The easy fix would have been to shut the barn door to the pasture. But I opted to inspect the fence by flashlight, so I unplugged the fence charger and walked the perimeter of the five acres. At the far corner I found the problem: The top fence line was down, probably knocked off by a deer leaping the fence but not quite clearing it. I tugged it back into place and secured it, then returned to the barn, where I plugged the charger into the wall socket once more. It clicked and flashed properly and I felt pleasure in the nighttime mission that took me around the pasture in the dark. At one point the fence would become my most time-consuming problem, the single truly troublesome aspect of the goat operation. But even in those first days I often walked the fence at night, by moonlight or sometimes in the rain, looking for a problem. Thunderstorms were a threat, because the expensive charger could be destroyed if the fence was hit by lightning. I'd hear a far-off rumble, often at two or three in the morning. Waking immediately, I'd wait for a second confirming crack, then run downstairs, throw on a raincoat, and rush to the barn to unplug the fence charger before lightning came closer. I never minded these forays into the ever-faster pelting rain, thunder booming close by; they lent an aura of adventure, and I'd return to sleep as quickly as I awakened.

Just when I was settling in and becoming more discriminating about what to worry over, I noted that the temperatures at night had dropped. Surely it wasn't cold enough yet for the does to start their circles of fertility. I called Susie.

"Yes indeed," she confirmed. "I've already separated my does from the billies. You'll want to do the same if you don't want kids in January."

So soon! It was not yet mid-August. I wasn't prepared. There was no place for the billy goat, no separate pasture and no shelter away from the does. The logistics seemed impossible, the problem difficult. I called John. It was a Saturday.

"I'll come by tomorrow after church and we'll figure out something," he said.

John is good at figuring out things; it's what makes his work fun, he often says. I had no idea how John would solve this one, especially on a Sunday when the lumberyard was closed.

Immediately after the Lutheran service, John came to the farm, work clothes on. First he took the metal T-posts he had used for the small temporary enclosure and pounded them into the ground within the pasture, marking off a small area for the billy. It included the wooden gate to the larger pasture so that I could enter without having to disconnect the electric fence and climb through the wires. Then he dismantled a length of snow fencing near the house. Together we wired it to the T-posts. Next John looked around the barns until he found enough scrap material to use for a three-sided shelter with roof, which he positioned so that the open side faced south. It was large enough for the billy and one or two other animals.

Then we dragged the reluctant billy goat to his new quarters and gave him some grain to mollify him in his estrangement from his does. The work took about four hours; the only thing left for me to do was to stain the new shelter black to match the barns and gates.

Shearing

"WE DON'T NEED anyone to shear the goats." Peter was emphatic over the telephone when we discussed the first shearing. "We'll shear them ourselves."

Shearing was the first real hurdle. It had to be done soon, sometime in September so that the goats could grow enough hair by the time cold weather struck. Six weeks' growth of hair was preferable, I was told. October is often relatively mild on the island because the sun-warmed waters surrounding it cool gradually in autumn, keeping the land temperatures higher than elsewhere. But by November the mercury can plunge into the single digits.

Despite Peter's insistence, I called around asking about shearers. It seemed that no one wanted to shear Angora goats. Their skin adheres loosely to their body, particularly around the neck, where it hangs in deep, baggy folds. It's easy to leave long, gaping wounds when shearing them. Furthermore, goats tend to collapse at your feet when you're trying to hold them in position. It's like trying to shear a sack of potatoes, a shearer once told me.

I called Donna's shearer. "No, I won't do Angora goats," he said first off. Then, "Well maybe, if you can't find anyone else. But I wouldn't be able to get to it until after Thanksgiving." Sheep are hardier creatures than Angora goats and can withstand the cold better; a November shearing wouldn't do for the goats. There was no one else, not a single shearer in the county. The shearers Susie Waterman used weren't available. The agricultural extension agent didn't know of anyone except Noel, Ingrid's husband. Both Ingrid and Noel attended the state shearing school one year and Noel occasionally sheared the odd sheep.

"At least we should get someone to show us how," I objected to Peter. After all, the shearers' refusals had indicated it was not easy.

"That's not necessary; we'll just do it. There isn't that much to it. Go ahead and order whatever equipment you need. Jeannine and I will be up the first weekend in September and we'll do it then."

The two-hundred-dollar shearing machine looked formidable, and the set of combs and cutting blades, used together in pairs, looked as though they could whir out of control in an instant. Accompanied by detailed instructions on its care, the machine also came with a booklet showing photographs of shearing, step by step. First, you place the animal on its rump. Then, with the shears, you shoot upward over the belly area, going against the grain with an even sweep. Never, never make second cuts, it warned. The hair won't be worth anything if you do. Once the underside is clean of hair, you hogtie the goat and shear the back in just a few sweeps from tail to head. That was the essence, with a few more refinements for shearing the head and neck.

Peter and Jeannine came to the island along with Anne and another child. The weather cooperated with warm, dry days and nights. Goats should be dry for shearing. The books recommend withholding water for twenty-four hours so that they won't drag their massive amounts of neck hair through their drinking water. And certainly they should be penned inside at night lest there's a sudden shower or even a heavy dew. We took all these precautions, assuming, I suppose, that we'd shear the whole flock that weekend, possibly that very Saturday. Peter glanced through the instruction booklet, gathered up the shearing equipment and inserted combs and cutters. We did not have the right heavy-weight oil on hand for lubricating the machine, something that's done frequently during the shearing to prevent it from overheating.

"We don't need it," Peter assured me. "The Three-in-One oil is fine."

Peter saw no need to bend over to shear, as was shown in the instruction booklet. He set up a makeshift table in the storage area of the goat barn. With the big door open to the outside, the space was airy and light enough to see well. We chose goat No. 200, or rather she chose us. A large, docile animal, she was one of the friendlier goats and did not shy away from us. My brother hefted her onto the table; Jeannine and I each grasped legs and tried to hold her steady. She looked stricken. We three were probably more tense than the goat, whom I later named Lucy after the Beatles' "Lucy in the Sky with Diamonds." Mouths set, foreheads furrowed, we set to work. It was grim, serious business. There was no lightness, no humor that day. Peter grasped the shearing machine; the cutter and comb whirred and vibrated back and forth. We kept the book open in front of us and began with Jeannine reading aloud from it.

Peter appeared perfectly confident, but his first sweep was

hardly more than a tentative pass along Lucy's belly. Then another over the same area. Lucy flinched. The comb must have been hot, possibly because of inadequate lubrication. I held my breath with every stroke; I'm sure Jeannine did, too. I also held my tongue about the second cuts.

It was painstakingly slow, the strokes hardly sweeps. But Peter wore his usual expression of mastery. Looking back, it wasn't at all bad for a first try.

"The book says no second cuts," I finally volunteered.

"Then why don't you do it?" he answered testily.

At the same time, Jeannine pointed out a couple of places where the hair had not been neatly removed. I must credit Peter, though, for not once nicking Lucy, who was by this time wild-eyed. As Peter said later, we three each had a different objective: He wanted to do the job quickly; Jeannine wanted the goats to look good after their haircuts; and I wanted to avoid wounds.

Leaving the neck and head untouched, Peter took off as much hair as he dared from the udder area to the brisket. It was slow going. It had already taken a good forty-five minutes. A professional would take not more than seven or eight minutes to shear a whole goat—and even less time for a sheep. At this point, Lucy seemed dazed; her mouth was open; she panted. It was hot; fat black flies buzzed around us. The tension deepened. Every so often Lucy stuck out her tongue and uttered a long, wrenching baa. I knew exactly how she felt.

It was time to tie her up to do the back and sides, and last, the head. Somewhere I had read not to use bailing twine because it could cut into the flesh, so I found a length of rope and hog-tied her, all four feet together as shown in the manual. Peter took up the shears again, first oiling them as directed. The broad, unimpeded area along her back and sides would have seemed easier had our tempers not been fast unraveling. The doe winced and jerked, her early calm dispelled by our inexpert touch. Jeannine continued to guide Peter from the book; occasionally I offered my interpretation. Finally the fleece was more off than on along the doe's back and sides. Her legs were a bit fuzzy, especially around the ankles, but most of her skin was cleanly pink without a hair to mar the look. It had taken almost two hours and we still had the hard part to do: the neck and head. We felt the heavy folds and deep creases around the neck. We looked at her ears and eyes and we three must have thought the same thing.

"How about if I take the hand shears and work on her neck?" I suggested.

"Go ahead."

Feeling much easier without the noise and power of the fast-vibrating machine, I hacked with the hand shears briefly, but they were dull. Lucy breathed heavily. She no longer resisted; she looked on the point of exhaustion, surely from nerves.

"Why don't we untie her and let her go?" one of us suggested. "We can finish her another time."

It was the obvious thing to do. No one had to suggest we quit for the day. That, too, was the obvious thing. Poor Lucy. When she stood up she looked like a badly clipped poodle: clean shaven along the back and sides, little ruffs at the ankles and a full neck and head of thick curls. But we were too tense to see the humor. We released her into the pasture with the other goats, then watched, horrified, as one by one they butted her and began chasing her around the field. It was enough that she had to endure our attempt at shearing; now she was persecuted by her own kind. Either the other goats did not recognize her or they reacted to our scent on her. I was appalled; I felt guilty. I'm sure Jeannine and Peter did, too. Soon friends of theirs stopped by, curious about the famous shearing. One look at Lucy and they doubled over in laughter. But at least by this time she was accepted by the others as a respectable member of the flock.

I felt thoroughly frazzled. Jeannine turned her attention to my small niece. Peter was grumpy, remarking that if one could divorce a sister, he would. Only Lucy recovered quickly. Peter conceded that a demonstration might be helpful, and I was instructed to call Ingrid and Noel and set up something for the following weekend. Peter would come up alone; we'd watch Ingrid and Noel shear a few, then we'd do the rest.

Ingrid and Noel agreed, and the demonstration was set for nine o'clock sharp on Saturday. Early that morning Peter arrived in good humor. Two guests were coming, and he's an enthusiastic and generous host, never in better form than when introducing visitors to the island and his beloved farm. Whatever hindrance the two men posed to the task of shearing would be more than compensated by an ambiance of geniality.

Noel is similar to my brother in that he exudes self-confidence, a master of all he undertakes. Ingrid is equally competent but more modest in aspect. Both are good-humored; I never think of Ingrid without seeing her smile and hearing her laugh. Noel ran through pointers: Always shear on a plywood sheet so that if you drop the shearing machine, it's less likely to break; adjust the tension on the machine frequently and oil it often so it won't overheat; separate the stained fleece from the clean right away so there will be no contamination of the good hair; don't make second cuts; be prepared to suture your goat in the event of a bad wound.

Peter brought up a stanchion he he'd had built at the workshop at his business. He intended to have the goat standing, elevated on its platform, head secured the way one might secure a dairy goat for milking. It wasn't the way Noel sheared. He took a goat by the horns, deftly set her on her rump, bent over the relaxed form, and sheared her underside.

"Be sure to put a finger on each teat," he cautioned, "so you won't risk cutting one off."

Once he finished the belly and chest area, he turned her somewhat so that she ended up in a ball at his feet. Bending down even more, he sheared her back and sides. Then, carefully, he did the loose folds of the neck and finally sheared around her head, putting a thumb over each eye as he maneuvered the heavy machine around the delicate bony contours of her face.

Noel sheared one more goat, then it was time for them to leave. It was also time for us to meet Peter's visitors at the ferry dock and take them to a late breakfast at the Sunset Resort. Later, looking out over white-crested waves rolling into the bay and eating Icelandic pancakes, we forgot the shearing. After breakfast there was a leisurely tour of the island before we resumed work.

"I'm not going to bend over the goats," Peter announced, pulling out the stanchion. "It's not necessary."

I held the next animal in place while he took up the shears. He looked good at it, like a professional. He wielded the heavy machine with aplomb, almost casually. He went down the animal's back from neck to tail and around its sides to the midpoint underneath, cutting with the growth of hair rather than the preferred way. The combs left little ridges of fleece but otherwise the hair seemed to come off well. Far be it from me to suggest that it was more properly done against the grain. Peter is one of those people who makes his own rules and generally succeeds at it; what does not work for other people capitulates under my brother's hand. The shearing still took a great deal of time but there was no bending, no turning the goat into an awkward ball at Peter's feet.

When a sheep is sheared, its wool comes off in one piece rather like a garment that is peeled off. Mohair falls from a goat in clumps. Under the hand of a good shearer, the fibers may hold together somewhat more. While Peter sheared, I tidied up, gathering stained mohair and stuffing it into a plastic bag, then sweeping up the good hair and putting it into a paper bag that I labeled with each goat's number so that I could weigh and record each fleece later. The hair was lovely to see fall away, because under the dingy gray surface, long curly locks were lustrous and creamy, almost silvery.

"May I try one?" I asked once Peter had finished a goat.

"No. Better let me do it."

I felt my resolve melt and my self-confidence diminish. He doesn't think I'll do it well enough, I told myself. And he probably thinks I'll take too much precious time.

Peter dispatched a second goat, his friends watching. He was steady and strong. His friends were impressed. I was impressed. And the more impressed I was by my brother, the more I doubted my own ability. Sibling competition.

After the second goat, it was time for lunch at the Sailor's Pub. Over potato soup and thick sandwiches, again shearing was all but forgotten except for the lingering scent of mohair that we were unable to scrub from our hands. Shearing ceased to be the focus of the weekend but rather was something to be fit in as there was time. After lunch we did two more goats.

It was my turn. One of Peter's friends remarked that I looked nervous. His comment didn't help. The machine felt heavier than I remembered; the noise was nerve-racking. What if the whirring cutter-comb combination flew apart? It seemed entirely possible, even likely.

I applied it to the goat for several inches, lost my nerve, and raised the cutting edge. Those easy clean sweeps in the book! I marveled at my brother's mastery. Tentatively I tried again. Above all, I didn't want to cut a goat. Peter was impatient. I tried again, achieving a longer stroke before losing my nerve.

"Here, better let me take it," Peter commanded, reaching for the shears.

I felt inept. Was Peter afraid I would injure the goat? Was he merely impatient because I was taking too long? I handed over the shears. Peter finished that goat and did one more. Then it was time for a walk around the farm and another short drive around the island before taking our visitors to the boat.

I've always thought newly shorn goats looked like skinny little old men in silky pajamas. They're sleek and silvery at first, with touches of pink where skin shows through the almost imperceptible stubble that's left. By the next day, the pink has given way to a white sheen of mohair. Shorn of their six-inch-long locks, the goats appear less than half their full-fleeced size and thin to almost bony with only wisps of chin hair left. Their sudden fragility and vulnerability elicits my protectiveness. Once I'm accustomed to their new look, I think of them not so much as skinny old men but as leggy young girls. Relieved of their heavy, itchy coats, they play and frisk about.

The next morning Peter sheared two more animals and finished Lucy before leaving. I didn't even suggest I take a turn. Time was too short for my halting efforts. All in all, with

Lucy it made nine goats. We had twelve more to shear and because Peter would not be able to come up again soon, I'd have to manage.

I called John. Yes, he'd help me. He'd like to, in fact. We'd do it the next Saturday, and he'd come around one evening the next week to take a look at the shearing machine and manual. Midweek as promised, over a bottle of beer at the kitchen table, John studied the instruction book for the shearing machine itself. He removed the comb and cutter, studied them and replaced them, adjusting their position. He turned on the machine and studied it to make sure the cutting surfaces were positioned correctly. He played with the tension, adjusting it to different degrees to see how the machine responded. When he felt comfortable with it, he asked for the shearing manual, which we studied together. I went over Noel's instructions and we discussed how we would proceed. I began to feel better about the job ahead of us. It would be easier with John. There are probably some things one should never attempt with a sibling.

Saturday dawned, sunny and dry. I've since learned to plan on rain or snow for shearing. The mere setting of a shearing date elicits a storm, I'm convinced. But that year nature was accommodating. We set up in the hay storage area, and we put sheets of plywood on the floor. Instead of using Peter's stanchion or tying the animals, one of us knelt and held the goat while the other one sheared.

John took the shears first. Like my brother, he's self-assured but his confidence is of a different texture than is my brother's. Impatience rumbles beneath Peter's steady exterior; John's calm affects me differently. But perhaps that's simply because he's not my brother.

After he sheared one goat, he handed me the machine. We would trade off. There was no suggestion I wasn't equal to the task; nothing indicated John was judging my performance. I was nervous. Mentally I gritted my teeth and started. John said not a word. He gave me no advice, didn't tell me how I could do it more efficiently. He let me figure it out for myself, which gave me the message that however I did it was probably as good as any other way. Upon reflection, I regretted having commented at all on Peter's second cuts—despite his secure facade, he probably needed approval, too.

The hair came off the goat's belly with surprising ease. The bony chest area was more problematic, but I got through it haltingly enough. With John holding the goat, I sheared the legs easily. With every stroke I felt myself relax a bit more. Last, I came to the neck and head. Suddenly I lost all sense of the animal beneath the hair. I had no idea of goat anatomy. With all those folds of skin I could see myself slitting the jugular vein and the goat bleeding to death in short order. And the eyes! What if I blinded a goat? The jaw and nose looked delicate and vulnerable. I applied the shears timidly for an inch or two, then raised them. I looked at John.

"You're doing fine, Susie," he replied to my silent request.

There wasn't much to do except keep on, bit by bit. After all, the object was not to have a beautiful goat, only a shorn one. To come away with a live and preferably unwounded animal was enough to ask at that point. If the hair happened to be cut well, so much the better.

We worked all day, John stopping occasionally to smoke his pipe. Patty came by and took pictures. Ann, John's sister, visited the barn. Other friends came by for a look or to help hold a goat. Everyone was cheerful and interested. Best of all, it felt like true teamwork. When we finished, some of the goats looked scruffy around the horns and face, but only one had a cut, which I dabbed with hydrogen peroxide. I was exhausted, from tension as much as anything, but elated to have the job done.

Later I sorted through all the separate bags of fleece, weighing each goat's hair and discarding any remaining stained hair. Our little goats were not yet full-grown and did not have the heavy fleece they would later grow. I had no idea what to expect but found myself rooting for particular animals, hoping they would have a good showing. Carmen's fleece came in at five pounds, Lucy's was five and three-eighths, Tosca's four and five-eighths Butterfly, a small goat, had a four-and-a-half-pound fleece. What would Mimi's be? I wondered. I hoped my friendliest goat, my favorite, would justify my affection by producing a luxuriantly heavy fleece that would make me proud. I would breed her confidently, assured that she would produce fine kids who would grow equally lavish mohair. We were not planning to cull our flock but rather raise as many goats as we could, so I did not need to justify a goat's position in our herd. But in my fantasies, I saw myself bestow my favor on an animal that was noteworthy in every way. I weighed her fleece last. Only three and a half pounds! I was dismayed. She's small, I thought; perhaps next shearing when she's grown a bit she'll surprise me.

The weights from our first shearing turned out to be light compared with later ones of comparably aged animals shorn after even fewer months of hair growth. These newcomers just up from Texas had been growing fleece since sometime in January, about nine months, and none had a fleece that weighed as much as six pounds. The difference we saw later—eight or nine pounds for a comparable animal—was due to better nutrition: rich alfalfa hay and grain compared with the variety of scrubby plants available for browse on the Texas range. But there is a tradeoff: The better the diet, particularly the more protein a goat consumes, the heavier the fleece and the coarser the hair. The northern goat raiser feeds her goats well, partly because of the availability of hay and grain and partly to help the animals withstand subzero winter temperatures. And as long as she sells her fleece in quantity by weight, she's in good stead. Poorly nourished goats produce lightweight fleeces of fine fiber. If the market consists of hand-spinners, who prize light, fine fiber, or if the mohair industry develops to the point that growers are penalized for coarse hair, the fleece of the poorly nourished goat may be more valuable than the heavier fleece of a well-fed goat. Currently mohair is classed roughly by the age of the animal, with a kid's first clip remaining the finest hair. With each clip the fiber diameter increases and the fleece is slightly more coarse. But all these concerns would come later, when we learned more about marketing our mohair.

Fleece

I'M WEIGHING and bagging fleece today. It's a job I enjoy and find soothing in the way that bookkeeping can be soothing. The weights I record don't deceive; they're not subject to whim or mood but rather tell me cleanly, without fuss, how the goats are faring. That's the logical part of the task; for the emotional part, I look over each fleece and exult at the shining luster of the hair from one of my favorite does or a particularly fine curl that's held despite the goat's age. The fleeces also tell me which goats may have health problems. A fleece that's unaccountably light or dry and dull is a warning to take a close look at the bearer. For today's task, I set myself up in a section of the barn where I've piled nearly one hundred bulging white plastic drawstring bags. These were labeled and stuffed with each goat's fleece as it came off the animal at shearing. Today I'll hang them from a scale to get a weight for each goat's mohair, then I'll stuff the fleeces into burlap feed sacks, grouping them by age and sex of goat. The chore takes all day these days, so I bring Molly, my older dog, to the barn to keep me company. She's unimpressed with my operation; goat smells are by now uninteresting to her and she feels no inclination to investigate the contents of the bags. For additional company, I set my portable tape player on an old wooden chair that serves as a work table and play cassettes of operas.

It's a serene day of work that I anticipate happily. It's in abrupt contrast to that first year after shearing, when my anxiety over the fleece haunted me. I had one hundred pounds of raw mohair and a bag of stained hair and not a clue as to what to do with any of it! Our whole project hinged on the fleece in the long run; for the short term, if I were ever to make any money during the first years of our effort, I'd have to figure out a way of turning each clip into something marketable. Peter seemed to have only vague plans for the future of the goat operation, and these were limited to the rather glorious outcome years hence. The intermediate part was gray and undeveloped and it seemed increasingly that it would be up to me to wrestle with it. The difficult thing was getting Peter's attention; after all, he had his business to run in Indiana and his life there. And, as he pointed out to me, the animals' mere presence on the land spreading their bacteria-rich manure would improve the soil that had long been dead in terms of microbes. The very enrichment of the soil of Peter's farm was becoming the raison d'être for the goats, or so it seemed to me. Whatever problems I had in figuring out what to do with the fleece he dismissed as merely part of the learning curve. We hardly spoke of the vision of a fiber-processing business that I fantasized might one day be my lifework.

The immediate problem was how best to separate the clean from the stained fleece. It was obvious with the badly stained, urine-soaked locks, which were brownish orange and reeking of goat. But what about the pale, honey-colored hair? It was not exactly the original creamy white of mohair, but on the other hand, it was not objectionable. I debated. I sorted, throwing honey-colored hair into a separate pile, then relenting and adding it to the clean, then changing my mind again. The difficulty is that the staining is permanent, although once washed the discolored hair turns into lovely odor-free fibers of rich earth tones: brown, ochre, burnt sienna, orange, tan, and beige. Beautiful as they are, they cannot take dye, and the superb way mohair takes color is one of its most prized properties. Later it was easy to tell which hair passed and which was truly stained. But in those first days even such a simple task cost much deliberation and uncertainty.

I spent the next months writing letters and making telephone calls, researching the possibilities of transforming our mohair into a moneymaking product. By telephone I met people in all phases of fiber work. I took notes of each conversation, then typed them for my file. I spoke with organic goat raisers in Idaho, with hand-spinners in Wisconsin and Minnesota, with small woolen mills in Michigan, Ohio, Wisconsin, and Illinois, and farther afield, in New England and Canada. I talked with people who had raised goats for years, and I called down to Texas and talked with the men who run the warehouses where mohair is sold on consignment to buyers from abroad.

I learned that hand-spinners love mohair and work with it in combination with very fine wool such as the fiber of the merino and rambouillet sheep. This is because mohair stretches, and a garment made of 100 percent mohair will "grow." I found that craftspeople use mohair in small amounts for Santa Claus beards and doll hair. But aside from a very small quantity sold to the occasional spinner or craftsperson, most mohair from goats in this country is sold abroad, primarily through San Angelo, Texas, the center of the mohair industry in the United States. It is shipped by the farmer or rancher to warehouses in and around San Angelo, where it is sold on consignment as raw, or unprocessed, mohair to buyers from England, Japan, and Italy. At one time the former Soviet Union bought quantities of Texas mohair because its warmth and durability made it valued for army uniforms. And I was told that mohair takes such hard wear that at one time it was used in the upholstery of seats in railroad cars.

I called some of the warehouses to inquire about the current price. Mohair is such a small commodity on the world market that the price fluctuates dramatically depending on the fashion industry and whether a fuzzy look in sweaters is popular. The price had indeed changed, from five dollars a pound when we decided to invest in goats down to sixty cents a pound! The price we could get that first autumn would not begin to pay the cost of shipping the fleece to Texas. There was no question of sending it there. And besides, we had planned all along to work with it ourselves.

I called commercial spinner after commercial spinner. It seemed that no one wanted to work with mohair, particularly a small quantity of it. No one, in fact, was set up to work with long fibers. Yes, some would take a little but it had to be mixed with wool at a proportion of 5 percent mohair preferably, and certainly not more than 10 or 15 percent. I read spinning magazines, searching for clues such as advertisements placed by commercial outfits that might take our fiber. I read in one magazine about a business in Canada that spun mohair using the worsted method, a process designed to use long fibers to their best advantage compared with the woolen method used by apparently all commercial spinners in this country.

I began collecting brochures from companies that might turn our hair into yarn. Before yarn can be made, the fibers must be washed, or scoured as they say in the business. Then they are carded and made into roving, which can be used by hand-spinners. Whatever we did with the fleece, whatever kind of commercial processing we elected—or eventually did ourselves—it would pare as much as five dollars a pound off the processing price if I washed the fleece first. I decided to try it. I had recipes for the operation:

—Take one pound of fleece;

— Put it in a canning kettle with a small amount of mild liquid detergent and cold water;

— Heat gradually to one hundred forty degrees, do not stir;

— Keep at a constant temperature for five minutes;

— Pour off and add more cold water;

— Repeat;

— Add detergent a second time and repeat the process;

— Rinse four or five times, heating cold water each time to one hundred forty degrees for five minutes.

I bought a canning kettle, enlisted a couple of other large kitchen pots and gave it a try. First I sorted through the mohair, taking out sticks, twigs, and bits of hay. Then I started the procedure, hovering over the kettle on the electric stove, fishing out any missed pieces of alfalfa as they rose to the surface. I kept an eye on a candy thermometer lest the temperature climb too high. The first wash water was black and gritty, but as I rinsed the fleece repeatedly the hair miraculously turned white. The job was unspeakably boring but required vigilance because too high a temperature would turn the fleece into felt. And it was arduous. The pots of water weighed more and more as I trundled them from sink to stove and back again. When the small lots of fleece seemed sufficiently clean, they had to be dried. Do not squeeze or wring, the instructions said. I carefully patted the hair between bath towels to blot the excess water. Soon the house was steamy from wet towels drying on the back of every available chair. I spread the first washed hair on a drying rack. When that filled up, I scrubbed the wooden floor in one of the spare bedrooms, covered it with newspaper and spread the damp fleece to dry. I worked for two days; in sixteen and a half hours I washed fifteen pounds of fleece.

During that autumn I washed sixty pounds altogether. I hated the job but was determined to make something of our fleece and could not think how else to go about it. The results were spectacular, at least to me. I was fast cultivating an appreciation for the silky, shiny fiber. But was it worth the effort? I was not at all certain.

My labor stopped abruptly in one of those fortuitous moments that masks as a problem. One day in December, just as I was preparing to launch into more fleece washing, I noticed water bubbling over the surface of the ground near the goat barn. It smelled foul, like a sewer. I called Lee, who is good at plumbing, then Lonnie, who pumps the septic holding tank. Apparently the septic field was saturated from the sudden onslaught of water rushing into it. The fact that the ground was frozen didn't help.

Peter later maintained that the oil from the fleece had plugged the pipes that carry wastewater into the septic field, but other people I met through fiber workshops said they had never experienced that sort of trouble in all their years of washing fleece. John assured me that septic fields become saturated over time, although Peter remained convinced that my foray into fleece washing had damaged the system irreversibly.

I was not daunted. Neither was Peter. He liked the idea of washing the fleece ourselves, or rather my washing it, and he was energized by the problem. He went to work on it and came up with a solution. He would design a system for washing the fleece out of doors over a wood fire. There would be a device to regulate the temperature; no hovering over the open fire would be needed. Furthermore, we could simply run the wastewater into the ground. It might take a couple of hours to get the fire up to speed, but once I did, I would have to add more firewood only three or four times a day, Peter claimed. I had seen the big cauldrons of maple sap boiling night and day over wood fires in late winter to make maple syrup. I envisioned something like that, which seemed far from foolproof. Better just to wait and see what happens, I thought. Fortunately nothing more came of it.

I was beginning to appreciate my profound lack of knowledge about the fiber industry. My natural inclination and concern for the project was in the care of the animals, and second in the shaping of a business. I love the colors and textures of beautifully woven garments but have never wanted to make them myself. However, I began to realize I should learn to spin in order to have a better understanding of the quality of fiber we would strive for as we developed our herd. It would be best to take a spinning course, I reasoned. Peter agreed, and I sent away for information about spinning wheels designed expressly for mohair. I could take the courses at Sievers, beginning and advanced spinning.

The spinning wheels were not cheap; they ran four hundred dollars apiece if purchased new.

"Why don't you order a dozen?" Peter was in an expansive mood. "We'll want to start people on the island spinning our mohair."

"But it's not that easy," I objected. "You can't just spin the mohair. It has to be washed and carded and heaven knows what else first."

"Then see about getting a carding machine and whatever else you need."

Peter's vision entailed having the Sievers School set up classes so that interested islanders could learn to spin our mohair. It was an attractive idea, but it takes time to learn to spin properly and more time to spin well. Was there really an interest in it? And what about the details of the project? Would we pay people to take the course? Would we pay them for their work by the piece or by the hour? Who would judge the quality of the yarn? Somehow it all seemed premature. We never had a chance to sit down and talk through our ideas; rather, they were thrown out in quick snippets of telephone conversations.

Although I was eager to develop a product, I didn't want to plunge in without spending some time gathering ideas, working them out and, in general, planning how we would proceed. And as the weeks went by, the kidding season loomed ahead of us, and that was a big enough unknown. It seemed a vastly better idea to concentrate on my animal husbandry skills, and at the very least, get through the first birthing. I did not order a dozen spinning wheels; it seemed foolhardy to run up a bill of almost five thousand dollars for spinning wheels with no one to use them. I didn't even order one.

I decided to test the mohair market in a small way. I sent most of the washed mohair to a commercial mill in Michigan to be carded and made into 100 percent mohair roving, the loosely twisted-together fiber that has been cleaned and carded and is ready to be spun. I planned to sell this at Sievers. If summer visitors to the island and hand-spinners attending Sievers' classes rushed to buy it, we'd order additional roving made of blended mohair and wool. It seemed a way to proceed, albeit a small, careful step.

Breeding

NOVEMBER, the twilight of the year, is normally a gray time on the island. October's reds and golds have been swept from the trees; December's pristine white is still to come. Most years it's sunless, a time to hibernate or travel mentally to other places, other times. But on the farm it's time to think about the immediate future, to decide which of the goats to breed and precisely when. Of course, the "when" is in great measure up to the goats, but with a little judicious arranging I can fairly well ensure that I'll celebrate new offspring of my favorite does sometime in early to mid-April. But typically, the practices I use now evolved from uninformed efforts at a solution. The question that loomed that first autumn was when to release William E. Goat from confinement so that he could at last fulfill the purpose for which he had been so carefully bred. Penned apart from the females since August, he had escaped only once. It was during a visit by a college classmate, Jean, and her family. One afternoon Jean noticed Bill chattering through the fence at one of the does. That night he made a break. The next morning the snow fence was partially trampled and Bill was on the other side mingling freely with the girls. I didn't notice any attempts at breeding then, but I could not be certain he had not mated with a few does during the night. I noted on my calendar that on or around January 18—approximately 150 days from Bill's night of freedom—we could expect a birth or two.

All September Bill was sweet and submissive with me, a model billy. He nibbled grain from my hand, pulled gently at my shirt when I came into his pen and bent his head down to have the base of his horns scratched. Surely, all talk about billy goats being aggressive, even dangerous, did not apply to Angoras. As the month wore on, he became increasingly excitable whenever females sidled up to his pen. By late October he was desperate. He tried to climb the snow fence; he flapped his tongue and snorted and pawed at his enclosure, and even attempted to mount does between the wooden slats. His thrusts were futile, his obvious frustration upsetting to see, and I was afraid he would injure himself. I had set the date for breeding—or more precisely for releasing Bill into the girls' area—at mid-November. Peter came up to the island one weekend and after watching Bill's antics, convinced me to move up the breeding date. I marked it on my calendar: October 28. We could expect kids starting approximately five months later. I told John, remarking that Peter felt sorry for Bill. He laughed. "Susie, if there's a male animal around, men will usually identify with it."

The optimum time to schedule the birth of kids is a subject of discussion and deliberation among goat ranchers; it varies with the climate and one's theories and preferences. The one constant is the gestation period, which runs from 145 to 150 days. Goats are seasonal breeders, with does generally coming into estrus and the males into rut in August, when the nights start getting cold and the days shorter. Breeding readiness peaks during October and November, and tapers off in

December. There are exceptions, however. One year we had two kids born in mid-December, and another year a kid arrived in July.

It's best to coordinate fall breeding with spring shearing, so that the does will be sheared three or four weeks before giving birth. Then when the newborns arrive, they will find a teat more easily and will not waste precious time—for the first hours are critical to maintaining their body temperature and gaining strength to develop—by sucking on a lock of hair. Also, shearing means extra hands to hold goats for inoculations, which I give several weeks before birthing so that the mother's immunity will transfer to her offspring. Many goat raisers like to schedule shearing and birthing early, claiming that kids born in February and March are larger and healthier than those born later. Certainly by the time for fall shearing, their fleeces will be heavier than those of late spring kids. But after experience with February births, I prefer to wait until April. In January, February, and even early March the island's temperatures can hover around zero or plunge below; fragile newborns must be moved immediately to a kidding pen warmed by a heat lamp. It means constant vigilance if the kids are to survive. I know at least two goat raisers who have full-time jobs away from their farms. When their kids were born early one year not one survived; they died of cold in the barn before anyone found them.

Whenever breeding is scheduled, the billy will be ready. There's no mistaking a billy in rut: He suddenly smells strong and rank and as the days go by, his face and legs become increasingly dirty. At first I thought Bill was merely rubbing his nose and cheeks against the black gate, but I couldn't account for his blackened legs. Then I noticed him repeatedly spraying himself with urine, stretching his neck so that his face as well as his front legs received a good dousing of a torrent of female-enticing hormones. All billies do this, I learned.

When released into a pen with females, a young billy in rut will rush around, joyfully attempting to mount any and all does at the first opportunity. Later, when he has had a chance to calm down, he will sniff the vulva of each female and sample her urine for signs of estrus. If one is ready, he will begin courting her, flapping his tongue and chortling. He'll smack his lips up and down her back, raising his head every so often to point his nose in the air and curl his upper lip to savor her scent. She will respond by shaking her tail and rubbing against him, although she will back away coyly, too. If she is not in estrus, she will run to escape him. He generally stays with her during her cycle, mounting repeatedly and quickly for only seconds at a time.

When I first reunited Bill with the girls, he raced around, apparently ecstatic at his good fortune and determined to capitalize on it as quickly as possible. The girls, though, did not appear to respond. I kept watching but saw nothing.

"Don't worry," Noel told my brother. "Sometimes they're shy and breed only at night. Besides, it takes only a few seconds. You might have missed it."

The one time I saw Bill make a serious mounting attempt that was not immediately rebuffed, he tripped and fell, landing with all four hooves in the air. Not a good sign, I thought. But my concern proved groundless; 147 days after the initial day of breeding, the first kids were born.

Donna's Kitchen

LADY MADONNA began to bulge around Christmas. When I rolled her onto her rump to trim her hooves, it was startlingly apparent. No one else was that large; next to her the other goats looked downright svelte despite their three months' growth of mohair. Could it be that she was bred when Bill escaped from his enclosure back in August? It seemed likely. The immediate task was to prepare myself with a cram course in goat midwifery.

I had not made time earlier because all December I had alternated between washing fleece and ferreting out information about mohair processing. The washing numbed my mind, but the research was energizing for it involved tracking down clues and fitting together pieces of a puzzle that I hoped would give me a picture of the mohair industry. However, it all began to look very unpromising despite a government subsidy on wool and mohair pegged at a price established periodically by the U.S. Department of Agriculture. The subsidy has since been phased out, but in those days a large-volume goat rancher, say in Texas, could expect a substantial check—one hundred thousand dollars or more—from the government despite the low market price of mohair. I was not even sure that our inexpertly sheared hair was salable. It began to look more and more as if there would be no money at all coming in from the mohair, at least not soon.

I was living solely on the rental of my condominium in Santa Fe and it was now vacant, the tenants having left suddenly at the end of the summer. There was no prospect soon for new renters and with no income, I had to choose between abandoning the goat project and selling my property. Leaving the goats to move back to Santa Fe when we had just begun the project was out of the question. I loved everything I was learning about them and took immense pleasure in working with them. And I was still convinced that the project had potential for a good income. The only solution, then, was to sell my house. I decided to make a trip to Santa Fe after the first of the year to arrange for listing the house for sale. Peter offered to drive me out; we'd make it a little vacation and take along my older niece and nephew, at that time in their late teens.

As the dark days of December wore on, I felt at an impasse. I was not discouraged about the goats in the long run and still had unbounded confidence in my brother and his dream. But I couldn't see how to move nearer our goals. Without articulating it to myself, the constant facing of worries about the goats and the realization of how inexperienced I was wore on me. I think that on a subconscious level I felt the weight of being alone with the responsibility, although it was not something I often dwelt on. The likelihood of Lady Madonna's early pregnancy with a due date during the coldest part of the year added to the money worries and the uncertainty about the goat project. At the very least, I had to prepare myself for early kidding, so I called Donna.

Seated in Donna's warm kitchen, steaming laundry hanging to dry near a huge woodstove, I felt enveloped in comfort. Donna made tea while I watched the antics of an orphan lamb prancing stiff-legged and less than steadily on the slick linoleum floor. Like so many other people on the island, Donna is quick to help. I hardly knew her then, but that afternoon she seemed like a longtime friend. She cut quickly to the essentials of my situation, the concern for our animals a natural bond between us. Over tea and pie we talked for two hours about the goats and sheep, about the mohair project and Peter's ideas, about my struggles with sorting and washing fleece—and the near impossibility of that task, particularly in winter when the washed mohair must be dried indoors. While I took notes, Donna listed supplies I should have on hand for kidding and sketched for me the course of a typical lamb birth, what to expect and what can go wrong. I came away feeling supported in my work; I felt there was someone I could turn to who not only could give me expert advice but who as a woman and a sheep raiser understood my struggles and my joys with the goats.

A few days later Donna came over to inspect Lady Madonna.

"It looks like it to me," Donna said, confirming a January due date. Together we rolled Lady Madonna onto her side and Donna demonstrated how to inoculate by subcutaneous injection for tetanus and enterotoxemia, known as overeating disease because it is associated with a starchy diet. Enterotoxemia is caused by the bacterium *Clostridia perfringens*, types C and D, which can cause sudden death in young kids. The mother is inoculated two to six weeks before her due date so that she can pass along immunity to her newborns; kids are inoculated at one month and two months for both tetanus and overeating disease, and thereafter on a yearly basis.

Donna advised keeping Lady Madonna in a separate pen inside the goat area while I was away in Santa Fe. A new mother might not know enough to come into the barn to have her baby, Donna cautioned, and a pen would also keep her safely out of the billy's way. Donna once had to shoot a ewe that was injured by a ram during her last stages of pregnancy. Eventually I would separate Bill from the does, but in my absence from the farm, I did not want to leave extra work in the form of hauling water and food to the billy's pen. Daniel, who works for our nearest neighbors, agreed to stay at the farm and care for the goats and dogs and cats. He would have enough to do, and with subzero weather ahead, I did not want to make his life even more difficult.

Despite money worries, there were occasional triumphs in those days, the kind of small flushes of success that come with each new challenge met. A heavy snow came in December that year. Until that time, getting in a shipment of feed from the co-op in Sturgeon Bay posed no problem. The grain came in 100-pound plastic bags and was shipped over on the ferry-boat. I simply drove the van down to the dock, loaded up with the help of the men at the ferry, then drove the four miles back to the farm, where I maneuvered the van across the grassy lawn and right up to the barn. I unloaded the bags onto a wheelbarrow to take inside and dump into plastic garbage bins. But with the heavy snow, it was impossible to drive to the barn. What I needed was a sled, which would not be difficult to acquire in most places in winter. But with only one general store, heavy on hardware and building fixtures, and one lumberyard, I might have to search catalogs to get one. But I was lucky; the Mercantile had a small plastic toboggan, for all of seven dollars. It quickly doubled for bringing firewood from the stack behind the barn to the wood box at the house, and later, when we had so many goats we had to store their supply of hay in a separate barn, I used it to transport hay, a bale at a time.

The other problem was the automatic waterer with electric heating element, which continued giving me mild shocks from time to time. The goats would not drink from it, so I bought canning kettles and hauled water from the hydrant in the barnyard. I called the manufacturer, who talked about stray voltage, something familiar to dairy farmers because their cows are occasionally shocked by the milking equipment despite proper connections and grounding.

"You mean electricity really is still a mystery?" I asked him.

"Yes it is," he said, and sent another heating element in case ours was malfunctioning.

Someone else suggested that the shocks were somehow connected to snow touching the fence line. This sent me out in a windstorm to try to hack deep, crusted snow away from the fence. It proved impossible, so I simply unplugged the fence charger but found the shocks still a problem.

Before I left for New Mexico John constructed a pen for Lady Madonna in the barn, where she would not be lonely, and together we clipped her back legs and belly area in preparation for the impending birth. Then, with everything in order and the waterer functioning again for no apparent reason, I left to meet my brother in Indiana and to drive to Santa Fe.

When I returned ten days later, Lady Madonna seemed glum and was not eating well. She was most likely depressed from the confinement, I reasoned. There was no sign of her udder ballooning—"bagging up," in farm parlance—which should happen within a week or two of giving birth, so I let her out of the pen for a few hours, which lifted her spirits. At that time, the billy's pen, unused since October, was too inundated with snow for occupancy, so he had to remain with the does and the pregnant Lady Madonna. She herself struggled so much when I dragged her back to her small enclosure that after a few days I decided to chance it and let her mingle with the other goats despite Bill's presence.

January 18, the first due date, came and went with no sign even of bagging up. January 22 passed; still no signs of teats enlarging or udder filling. If I had just once seen a pregnant goat ready to give birth, I would have known with reasonable certainty that Lady Madonna was not at the point of delivery. By the end of January, it seemed that she was indeed pregnant—but not from an August conception.

First Birthing

How MANY goats had conceived and were actually pregnant? By March the question teased me constantly. It accompanied me to the goat barn where I studied the animals, searching for confirmation. I watched the does, shaggy and rounded with nearly six months' growth of fleece, and I'd speculate which ones surely were of a girth caused by pregnancy and not just an abundance of hair. In the evening after feeding them, I'd lean on the open Dutch door looking at them. Sometimes to my eye, they'd all bulge; other times I wasn't so sure. Lady Madonna, certainly. The rest? It was anyone's guess. Peter asked a psychic; she predicted a number, less than a dozen. Susan Drummond, in her book *Raising Angora Goats the Northern Way*, wrote that of her first Texas flock of twenty does there were only four births, and she attributed the paltry showing to their recent move to the north. Ours were on an almost identical timetable; I didn't want to get my hopes up.

By that time I felt I had mastered hoof trimming. In three hours' time I could catch each goat, work her to the ground by gently pulling a hind leg to tip her off balance, then cradling her against my knee, I'd trim. First I'd gouge out the manure packed into each toe so that I could see the horny walls, then I'd cut away the excess outer layer of hoof. I was determined that the goats not be subject to undue stress, so I lured them to me with handfuls of grain, and avoided chasing all but the most bashful. I reduced my earlier hoof trimming by about five hours.

I wasn't so confident with injections. In early March John helped me give shots of Ivomec, a new wormer, along with inoculations for tetanus and overeating disease. I gathered some needles—I had not yet learned to buy them in boxes of one hundred—rubbing alcohol, cotton balls, syringes, and a list of goats. It was important to administer the shots well before kidding, and because it was the first time for our Texas goats to be inoculated, they would need a follow-up injection in a few weeks. John and I took turns holding goats and injecting the two substances, the wormer and the combined overeating disease and tetanus toxoid. We poked the needle under the skin at the side of the neck near the top of the shoulder. Were we inserting the needle too deeply? What if we struck a nerve along the spine? The goats were easy— patient and unflinching. John was calm and if he thought my worries silly, he never let on.

That night I watched the goats closely. Were they a tad lethargic? Could I have overdosed them? It seemed entirely possible; after all, the amount of wormer is based on body weight, which we had no means of determining. Would I come out to the barn the next morning to find them all dead or dying? I also worried about the dose of toxoid, which was set at two cc's per sheep, which are generally much larger than Angora goats. The veterinarians assured me it was appropriate, but I knew they had little or no experience with Angora goats. Did they even realize that these animals were so much smaller than sheep and dairy goats?

The next morning the goats still seemed sluggish. It's my imagination, I told myself, and it probably was. However, I did not suspect that abscesses were forming at the injection site on some of the animals. When these became noticeable a week or so later, I called the doctors. Goats are notoriously subject to abscess, but at least these were clearly due to the injections rather than to the contagious bacteria that can run through herds causing internal abscess and death. Possibly they were caused by the inoculant itself, but more likely by a less-than-clean needle despite my precautions.

"Try to drain them if they are ready, then squirt some hydrogen peroxide into them," the vet advised. And next time, he added, we might try giving the tetanus and overeating vaccines separately.

That first year was a time of feeling my way, of not knowing which aspects of the goats to attend to, of not yet being able to read the signs they gave me. However eloquent the goats were and however eager I was, we failed to communicate much of the time. I'd worry about inconsequential things and fail to note the important ones. For example, I had not observed any mating activity, and not wanting to be disappointed told myself it was unlikely that more than a few goats had been bred that first year. Consequently I had only a vague idea of when to expect kids. Probably in early April, I thought. I was quite certain there had been no breeding when Bill was initially released into the girls' pen. I didn't see any and furthermore, we never introduced a teaser buck to get the does in the mood. The books say that the does cycle in response to the presence of a billy, so to get them cycling right away goat ranchers often put a sterilized buck among them so that the billy will not exhaust himself chasing them around before they are in estrus.

The important thing, I knew, was to plan for shearing in advance of birthing, sometime around mid-March. I hoped Peter would come up to take part, but he said he would be busy elsewhere. And there was a goat-raising seminar arranged by Susan Waterman with a grant from the state. Peter and Jeannine and I planned to attend, so I'd have to work the shearing around it. John reluctantly agreed to help again. He was not enthusiastic about it; it had been hard enough on his back the first time. We'd have to schedule it on a weekend, he said, and the first one he was free was after the seminar.

Jeannine, Peter, and I met in Madison for the two-day workshop for prospective Angora goat raisers and novices. An expert breeder and livestock judge from Texas gave lectures and demonstrations on various topics, from nutrition and parasites to the difference between raising goats in the north and raising them on the Texas range. We practiced scoring fleece for fineness, crimp, and luster; we learned how to spot kemp, those hollow, wiry hairs that will not take dye; we evaluated billies for hair covering, stance, straightness of back and overall proportion of body. The group ended at Susie Waterman's for a discussion of the merits of her prize breeding males.

Peter had been called away on business but returned in time for the afternoon session at Odyssey Farm. He didn't go through the exercises of grading goats with the rest of us. While Jeannine and I dutifully patted goats, ran our fingers through their hair, studied them, fretted when we couldn't locate a strand of kemp, deliberated on coarseness and luster and crimp, and scored our sheets as we were instructed, Peter simply looked over the registered males and picked out Susie's best billy to add to our herd. Susie would deliver him in the summer.

"Four hundred wethers," Peter had announced earlier in the year, "and enough does to maintain the number." I'd taken this statement as a joke, but I don't think it was. At least not quite. He concluded his business with Susie by arranging to have her go to Michigan to buy twenty more commercial-quality does, which she would bring up to Peter's farm with the new billy.

We were already at capacity in the barn, I pointed out.

"Don't worry," Peter said. "I've already planned an addition. I've sketched out how I want it. John will build it."

"But when?" I asked. I'm not sure I received an answer.

That Saturday John and I sheared. We blanketed the cement floor in the barn with plywood sheets, and oiled the shearing machine. I dragged out the first goat, Mimi. As the hair fell away from her belly, I looked for signs of pregnancy. There was no indication that I could see. She looked well fed. But pregnant? I thought not, based on the lack of any definition to her udder.

As in the fall, I was nervous when I first turned on the shearing machine but felt better after a few successful sweeps up the belly area. John's calm again had a steadying effect and as in the fall shearing, we took turns either holding a goat or using the machine.

"Look at this, John! She looks pregnant to me," I exclaimed.

The goat after Mimi indeed had a large belly with her skin stretched taut across it, which made the shearing easier. But what I noticed was not her ample size as much as her udder, which was developing into a vessel or pouch that was only sparsely covered with hair. Also, her two teats were larger than the floppy little one-and-a-half-inch appendages that one normally sees on a female that has never had kids.

"I'll mark a P next to her number," I said cheerfully.

I was excited. We'd have at least one kid. I didn't realize then that if these goats had been veteran mothers, their udders would already be enlarging very noticeably. Two weeks or so before kidding, the udder in a goat who has already had kids will become increasingly pronounced, or bag up. By the time the goat goes into labor, it looks like a large, taut balloon dangling between her legs. Normally, though, with first-time mothers, the udder swells only just before they give birth.

"Look at this one, John. I'm putting two Ps after her number."

We were shearing the next one, whose udder was even more defined. It was exciting and added a new dimension to our work. I forgot my fear of cutting a goat and my anxiety in my eagerness to assess the pregnancies of the animals we worked on. From seven-thirty in the morning until we stopped at three-fifteen in the afternoon we sheared thirteen goats. By that time I had a list with seven does with one, two, or three Ps after their names.

The next day was Palm Sunday, March 24. My closest neighbors, Barbara and Ray, had invited me to a festive family noontime dinner at their house. John and I would resume shearing after the meal, finishing up the goats then.

The phone rang at Barbara's just as dessert was being served. It was John.

"Susie, you'd better get over here fast. One goat's been born and another is trying to be."

I raced to the van and drove home. Some of the lunch guests with small children followed in their cars to see the baby goats. At home I pulled on jeans and rushed to the barn.

John had already stacked hay bales to make a kidding pen in the shearing area. Much earlier Peter had brought up four-by-four-foot hardboard panels for kidding pens, but they were stored in the utility room, buried under miscellaneous equipment. Only the day before I had made a mental note to extricate them the first thing after shearing.

The newborn kid was already up and wobbly next to its mother, a goat who had not yet been shorn. John led her into her new pen while I carried the baby. The kid seemed dry enough; I wouldn't need to dry it further with paper towels, something the book recommended doing. Mentally I ticked off the instructions I'd read. The baby should be nursing within minutes of birth. It was critical that it take nourishment immediately so that its body temperature would remain normal. The first milk, besides, is the colostrum, which flows for about twelve hours and gives the infant protection against diseases during the initial weeks of life. Sometimes the babies have to be held up to the teats and a little milk squirted on their noses and the outside of their mouths to get them to drink. Already I was concerned that the new baby was not yet suckling, particularly because I did not know how much time had elapsed since his birth.

I turned my attention to the other baby. I looked for a moment into the does' enclosure, trying to find a mother in childbirth. If she had been in labor at one time, she wasn't any longer, it seemed. Then I saw it. A sheared doe was running excitedly about and under her tail bobbed a little purplish-red head. Something was wrong. A baby should emerge front hooves first, nose following closely behind. A second glance and I knew I'd need help. I raced back to the house and called Donna, left a message on her answering machine and called Ingrid.

Donna arrived first, followed by Ingrid and her daughter.

Donna assessed the situation. "We won't be able to get it out the way it is. We'll have to push the head back in so that we can work the legs out."

I caught the doe, Gilda, and we held her in the hay on the floor of the goat pen. The baby was large; the doe, still not fully grown, was small. Possibly there had not been space for both legs and head to emerge through the birth canal. Donna greased her hands with cooking oil and pushed the baby's head back into the doe. The baby looked dead. Gilda bellowed and struggled; Ingrid and I held her as firmly as possible. Donna inserted a finger in the narrow opening to try to find a foreleg to bring upward and out. She worked her hand in but could feel nothing at first. Then Ingrid gave a try while Donna and I hung onto the goat. Our efforts seemed to take hours, although only minutes had passed. Every so often I called in to John to ask him if the other baby was nursing. Plunging her hand deep inside the goat, Donna felt around and located a leg pressed back against the kid's side. She hooked it with a finger and worked it forward until it came out. We would need both forelegs forward and out before the baby could pass through the opening, Donna explained. Since then, I have managed to birth kids with only one leg forward when time was running out, but they were small kids and came out more easily than this baby. At this point I think we knew the baby was beyond reviving. The doe continued bellowing, probably in pain from Donna's hand and wrist inserted well inside her. A good half hour passed before the second leg and then the baby itself came out. Donna grasped it by the back legs and swung it around to clear its mouth and lungs of fluid. When she put it down, she bent over and blew into its mouth, trying to start it breathing. Its little chest inflated, then sank. It was no use. There was no heartbeat. Probably it was dead at the time John first noticed it.

"Such a very large baby," Donna observed. "That was the problem. It couldn't pass through the birth canal the right way."

I penned Gilda and gave her the penicillin shots that are recommended to prevent an infection from bacteria introduced by putting a hand inside the goat. The crowd of small children and mothers had melted away by this time. I turned to the live newborn and tied a string around the wet, stringy remains of his umbilical cord, clipped it beneath the knot and dipped it all in iodine, just as the book instructed. Next I checked the doe's teats for the waxy plug that the book said must be scraped off with one's thumbnail. It didn't seem to be there; either it had not formed or the doe had bitten it off herself. The new mom, Leonora, was not making things easy. She stood her ground, her baby at her feet, and snorted at John and me. I worried that the baby wasn't drinking, but it seemed healthy. In the first few hours, even days, if a baby is not getting milk, the inside of its mouth will be cold. I stuck a finger in its mouth; it was warm.

By the time we finished it was late afternoon. I couldn't face shearing. It would be too stressful for the goats, I reasoned. I knew it would be too stressful for me.

The rude, bright sound of the alarm clock jolted me awake at two in the morning. Drummond's book advised checking the barn every three hours during kidding. It also cautioned that minutes after you leave the barn a goat may go into labor. After the day's introduction to life and death at birthing, I was determined to check the goats every two hours round the clock. I pulled a coat over my nightgown and robe—by the second night I slept in my running sweats—stepped into rubber boots, grabbed a flashlight, and trudged out to the barn.

A persistent raw wind clawed at the weathered boards of the barn, but Leonora and her baby seemed safe from drafts, nestled down as they were in thick hay behind bales of alfalfa. A kid is susceptible to chilling because its body cannot effectively regulate its temperature at first. I reached down and stuck a finger in its mouth; Leonora snorted her threats. The mouth was warm. It must be drinking, I decided. In the next pen Gilda was awake and quiet but had not eaten the grain I'd put out for her.

I hesitated at the door to the pen where the rest of the goats slept. What if there was another birth gone wrong? What else might I find and have to cope with alone in the middle of the night? I listened at the door; there were no sounds except the soft, rhythmic grunting I've since come to associate with pregnant does. I switched on the light and looked around the room. In the far corner, something white and flat lay in the hay. A baby, surely. Was it alive? I rushed in and glanced around. The mother must be Louise. She was the only one with a discharge dripping down her back legs. Unconcerned, she stood nibbling hay with her back to the baby. I scooped up the kid; it was cold but still alive. I walked back to the house with it, picking my way carefully over icy banks of snow. Once in the house I shooed the dogs upstairs and shut the door to the kitchen. I grabbed a bath towel from the back of the linen closet where the old towels are folded, wrapped the infant, and turned on the oven to produce heat quickly. Thumbing through Drummond's book to find what I had read ten times already—like a new mother with Dr. Spock—I sat holding the kid near the open door of the oven to warm it. Slowly, as the kid warmed, it seemed to take up the life that was temporarily postponed in the barn. Its mouth, though, was cold; it was not out of danger and didn't seem ready to drink on its own. With the chill dispelled I could safely inject a 50 percent dextrose solution to give it a little sugar for energy. I put the bundle in a box, then fumbled for the bottle of solution and a syringe. Drummond advises giving it in several places subcutaneously, under the kid's legs. I did this nervously, with trepidation. The kid, a little female as I finally noticed, responded. So far so good, I thought.

Next I took a glass from the cupboard and returned to the barn to take some milk from the mother. Louise was never one to oblige by being friendly or even cooperative. I cornered her and wedged her against the wall with my shoulder; then, steadying her with one hand, I tried to milk her with the other into the glass I placed under her teats. I'd never milked anything before, had never even seen a demonstration. I grasped the small appendage and ran my thumb and two fingers up and down. Nothing came out. I tried again, starting higher at the udder. At the same time, I was conscious that time was running out for the baby. Finally a drop oozed out and into the glass. The goat jerked; the glass turned over in the hay. I let her go to search for a relatively clean pail, then attempted it again. The minutes seemed long. I had no idea how much time the baby had. I'd heard that on the Texas range between 50 and 60 percent of newborns die, victims of maternal neglect or chilling rain and wind. The thick colostrum finally began to dribble out, not copiously but enough for a start. Back at the house I drew the milk into a syringe and dripped it into the pocket between the baby's lips and gums. Magically, at the first taste, the baby began to meet life. She looked for more milk and I gave her the small amount I had, then left her once more in order to construct a makeshift hay-bale pen in the barn. It was frigid at that hour so I decided to rig up a heating lamp. A quick search of the utility room turned up a long board, which I threaded through the rafters. I hung a lamp from it, securing the lamp with wire and letting it dangle about a foot from the top of the pen. Then I plugged it in using two extension cords from the house. I settled Louise in the pen with grain and water, then retrieved the baby and held my breath. Would the mom accept her kid? There is always a danger that a mother will not accept her newborn after a separation. But this time the doe grunted softly and reached around to begin licking the baby ever so tentatively. Standing unsteadily, the baby began poking her mother, looking for more milk. Better to let them become acquainted on their own, I thought.

I tore myself away from the barn, excitement welling in me. Awed at seeing a cold, nearly lifeless kid revive and in an hour's time stand and drink, I felt I had just witnessed something powerful and mysterious yet elemental. At the same time I felt an instinctual side of myself move into place, as if for the first time I claimed a facet of life that I had not had the opportunity to develop.

There was no question of going back to bed, although it was only four-thirty in the morning. Too exhilarated by birthing and adrenaline to sit quietly, I began cleaning the house. I could hardly wait until a reasonable hour to call Peter and Jeannine and let them know. Throughout the morning I checked the goats again and again and finally stopped, suddenly exhausted, only to revive instantly upon entering the barn again.

Most of the day I fussed over the goats, repeatedly poking a finger in the mouths of the two kids to make sure they were warm. My mind was riveted to the goat barn in those days; there was no room in my thoughts for anything else. It seemed that I was taking a crash course in animal husbandry, hands-on.

John came that afternoon to finish the shearing. We cleared a spot among the makeshift kidding pens and worked into the evening, finishing the last goats easily. Suddenly shearing seemed effortless in comparison to this new world of births and deaths. After the last clean-shaven goat scampered away, John and I leaned on the door to the pen surveying our work. Watching animals is a part of livestock raising. It's an information-gathering time when the farmer lets her eyes wander over the animals, her mind open and receptive to whatever they may tell her.

"Look over there, Susie. It looks like that goat's water has broken," John pointed out calmly.

It was Lucy, the goat that Peter, Jeannine, and I had first tried shearing. John and I went to work. John stacked bales to make a pen; I led Lucy into what had become the nursery and maternity area. Once in her new pen, she pawed the hay, then half lay in it. Soon she pulled herself to her feet again and repeated the pawing as if fashioning a nest and testing it to see if it suited. John and I watched. Surely if the water broke, the baby would come soon. I now know that water bags, full of amniotic fluid, mark the birthing site for the doe. The smell of the fluid seems to trigger her reactions and induces her to stay at that spot until the kid is born. If she is moved away from the wet spot, she may seem disoriented until the kid comes and she smells it.

Within minutes John and I saw something protruding from the goat's vulva. Contained by the sac, it appeared surrounded by jelly and was difficult to make out. Then I realized just what we were seeing: a snout with tongue lolling out to the side. The legs should have been first, hooves with nose tucked just behind.

"John, we've got to do something." I was grateful I wasn't alone.

Did we go back to the house to scrub our hands? I don't remember. I hope so. We greased up with the bottle of cooking oil that was still by the barn door from the day before. We got Lucy down on her side and I slipped a finger into her, alongside the little head that was trying to come out. The warmth inside the goat surprised me. There seemed no room for more than a finger, no room to search for a leg. John and I took turns trying. We pushed the little head back into the doe slightly, which gave us more space to maneuver. The trick was to locate a leg and bring it forward. John found it first. We took turns in the warm, slippery cavity, trying to bring the leg up, but each time it slipped back. Suddenly John succeeded. We still had to reach the other one and we worked for what seemed too long. Would the baby suffocate while we fished for its leg? Would it drown in the birthing fluid? I asked myself these things while running to the house to call Ingrid, who said she'd come immediately.

When I got back to the barn John had just taken a piece of tissue to hold around the second hoof he'd located. With a better grip, he pulled it forward and with it, the baby slithered out. We looked at each other for a moment, both of us kneeling in the hay beside the goat, then hugged in relief and jubilation. I felt tears running down my face. Lucy did not react but rather lay there; she seemed exhausted. I picked up the wet, struggling kid, male and very much alive, and placed him at her nose. Instantly she, too, came to life, and as she did, she began to nuzzle him and lick him and utter her goat lullaby of soft, cooing grunts.

First Birthing, Part Two

Time began to lose meaning. The hours in the goat barn sped by, but the hands on the clock pointed to numbers that had no relevance except to mark intervals. I no longer felt the sharp, bitter night air; elation mingled with anxiety eclipsed the cold temperatures, and the anxiety was the dread that I might not be able to handle whatever I found in the barn. I hovered over Louise and her baby, who seemed weak. Lucy's large baby was a dramatic contrast—vigorous and eager to drink from the very first. I gave milk to Louise's kid from a syringe and fussed over the goats until four-thirty in the morning, then set the alarm for six o'clock and slept.

I approached the barn just as the sky grew light, and as I did, I heard muffled catlike cries. I rushed in and saw Lady Madonna in a far corner nuzzling one baby while another lay close by. I did not stop to inspect more carefully but stepped back and began pulling and shoving bales of stacked hay to make another kidding pen. Only when the new pen was ready did I grab a towel and go to the babies. They seemed particularly weak and small, but I was no expert. I moved them and their mother into the pen and went through the routine of dipping the cord in iodine and scraping away the plug from the mom's teats. I held each baby boy up to a nipple and squirted milk onto his mouth, then watched as Lady Madonna nuzzled and licked, grunting softly all the time. Because it was cold and there were two, I used a towel to help dry them. The mother was doing all the right things but even in my ignorance I sensed something wrong. I kept checking their mouths; they were warm. No cause for worry, I tried to convince myself. I turned to Louise and her baby. Again I couldn't quite articulate the problem but they, too, concerned me.

By eight o'clock the twins still did not stand; consequently they were not yet nursing. Time was running out; already they had lost two hours of the precious twelve hours of colostrum flow. Lady Madonna was still exemplary in her attentiveness but I was afraid that even such a diligent mother would lose interest in babies that failed to take her milk. At the same time I noticed that Louise had not yet eaten the grain I offered her the day before, whereas the others had gobbled hungrily from the first moments after birthing. I needed to attend to her but also to the newborn twins if they were to survive. Whom could I call to help? I wondered. I thought for a moment, then remembered that Melanie, an acquaintance, was studying to be a midwife's assistant. She'd be perfect. I called her and yes, she'd be over as soon as her children left for school.

While I waited for Melanie I turned to Louise. Studying her, I decided to try Pedialyte, a solution for human infants to restore the electrolyte balance when they are vomiting and have diarrhea. I drove to the store and back, then plied the resistant doe with some of the cherry-flavored liquid, which I administered by turkey baster. Next, a shot of selenium. This would be my first solo injection into an adult animal, and the doe had no intention of cooperating. I drew one cc into a syringe and grasped her leg. She would have none of it. She pulled away and swung her weight around so that she faced me. The baby huddled in a corner, clearly not safe from the mother's hooves. I tried again. I leaned into her with my shoulder and managed to insert the needle into the muscle of a hind leg. She lunged forward, taking the needle with her. The syringe remained in my hand. Instantly the needle became an instrument of harm. I sprang at her and managed to withdraw it without either of us squashing the kid. Louise snorted and drew her hind leg up and refused to stand on it. Good Lord, I thought, I've hit a nerve. Probably I've injured her for life. But after a minute or two she put her leg down and stood normally.

By that time Melanie arrived. Moments later we were each settled on a bale of hay in the cold, drafty building, each with a twin enfolded in a bath towel. I milked colostrum from Lady Madonna; carefully we dribbled milk around the kids' mouths with a syringe and gently rubbed them. The twins hung limply at first, legs dangling, heads drooping to the side. Melanie didn't seem fazed. She cheerfully kept at the task, chatting with me as we worked with the babies. Patiently, gently, she coaxed the babies along, talking to them, dripping milk into them, massaging them. Slowly their muscles tightened, their limbs took strength, and they came to life. Two hours later they stood shakily and nursed on their own.

I learned from Melanie that morning that I mustn't expect results instantly from any given procedure. And I learned that patience must prevail in the face of the vagaries of raising any living creature. In a single morning I began to appreciate what ten years of motherhood, common sense, and a strong maternal instinct had taught Melanie. At least I had been toughened for this work by my newspaper job, I thought. Working three or four nights in a row until one or two in the morning at the paper cultivated my endurance. I was hardened to the adrenaline rushes of deadlines, to the exhilaration, the exhaustion, the years of irregular sleep, the sensations of anxiety followed by the glow of accomplishment. Kidding was little different.

John came by at noon.

"Susie, why don't you take a nap and let me watch the goats while I straighten up the barn?" he suggested. The offer felt like a gift. I stretched out on the bed, giving in to that delicious sensation of tiredness when every bone sinks into the soft fabric of the mattress.

Some of us crave sleep when tired, but even more, some people need order. If my world is calm and neat to a degree set by some internal measure, I function reasonably well without a night's sleep. But add disorder and confusion on top of sleep deprivation and I frazzle. The goat barn was chaos, and only the goats appeared serene. The maternity pens were makeshift; hay bales, once stacked neatly like cans on a pantry shelf, were pulled askew, dragged apart to construct pens and pushed aside to create space for more pens. Paper bags of fleece from the shearing, stashed in every corner, threatened to disgorge dirty mohair onto a filthy floor. Shearing equipment and the untidy remains of that operation had been simply shoved to one side. Iodine-stained towels for drying kids were draped over sacks of grain, and there was a jumble of heating lamps and extension cords and various other farm items that had little to do with the work at hand. I slept despite this.

When I returned to the barn in two hours' time, John had transformed it. All the numbered paper bags bulging to overflowing with fleece were neatly stacked on a long high shelf in the adjoining room. He had swept away the debris of shearing and had hauled out the hardboard kidding pens and set them up and moved the goats into them. He had restacked all the hay bales, moving some out to provide maximum maternity space. Inside the actual goat pen, John had rigged up a watering device, attaching it by hose to the outside hydrant, so that I could pen the animals at night but still allow them access to water. In a short two hours, John had banished confusion and produced order and along with it, the gift of serenity.

I was sometimes aware of being alone during that first kidding season. I missed someone to share the joy and excitement. And I missed having someone to turn to merely to inquire, "What do you think?" But I wasn't always alone. John was frequently at the farm working on carpentry jobs for my brother, and I knew he had a genuine interest in the animals. Often at the end of the day we would lean against the fence watching the goats, John pointing out when an unborn baby kicked against its mother's side. Goat watching, or perhaps any sort of animal watching, is a pleasure to share; John often provided that companionship. For the big problems, I could always call the vet in Sturgeon Bay or Susie Waterman or Donna or Ingrid. I muddled through, hoping that my judgment was intelligent, that my instincts were sharpening, that I would not fail the goats.

During that first kidding season John made a habit of telephoning every evening to see if we had any new births. Mainly, I think it was an excuse to see if I needed help. After the first difficult delivery and death of the kid, I knew there would be births that could not be handled alone. Donna and Ingrid both had children at home and I didn't want to trouble them in the middle of the night. Just as I wondered whom to call in an emergency, John volunteered to help at any time of the day or night and I knew he would do it gladly.

After Lady Madonna's twins there was a two-day break until the next birth. However, because I had no idea when to expect the next births, there was no feeling of respite. I kept checking the barn every hour and a half, expecting to find a newborn or a doe in labor. Now I breed the goats selectively and have a better eye for impending labor. As I became more confident during those first days, my trips to the barn in the hush and blackness of early morning felt like childhood Easter mornings, the surprise of discovering a pure white kid in a nest of hay very like finding those colored eggs magically in unexpected places. Perhaps it was the anticipation of the miraculous in the quiet of early morning that linked the present with childhood.

The weather on the island in late March and early April is cold and stormy. I've never known it to be truly mild or the least springlike. That April there were snowstorms and howling winds that left huge drifts circling the goat barn. I climbed over hip-high banks of icy, packed snow or through softer, newly drifted mounds. I seldom feel the cold upon awakening, and in the barn in those early morning hours I would remark to myself that it really wasn't that cold after all. Then I'd notice the water frozen in the pails in the kidding pens. One morning around three-thirty a west wind started up and blew furiously. The walls of the goat area itself had been covered so that despite cracks in the weathered exterior boards, it was fairly free of drafts. But the maternity area had not been improved, used as it was merely for storage. Bales of hay stacked along the south wall insulated that side; the east side was the goat pen and the north was simply an interior wall with a utility area on the other side. The west side remained unprotected; raw, icy wind rattled the boards as it swept into the barn now crowded with kidding pens. The immediate solution was to move the hay bales and restack them along the west wall. Later, in the first light of morning, I tacked odd pieces of wood and fiberboard on the outside of the barn to cover the gaps and holes.

Eleanor Rigby's male kid came just before midnight on Friday. He stood and suckled almost immediately, crying his loud infant cry, rather like that of an insistent Siamese cat. The strident cry helps focus the attention of the mother during the first hours, but he kept crying after most newborns lose their loud voice. He drank, though, and his mouth was warm, so all seemed well.

On Saturday Carmen's large udder suggested she would give birth soon. I pulled her into a kidding pen in the afternoon, then set the alarm for four o'clock and took a short nap. When I checked the barn shortly after four, Carmen was licking her baby, a little male. I attended to the two; the baby was already struggling to his feet, trying to find a source of milk. I marveled at the instinctive efforts to survive; to be born knowing what to do first is almost beyond comprehension. I was pleased with myself, too, for recognizing that the birth was at hand and putting Carmen in a kidding pen. However, I learned later from experience that most does will not tolerate a kidding pen until after the baby is born.

I glanced around at the other goats, then looked outside. It was a gray day, cold and raw but not frigid enough to keep the recently shorn animals indoors. Some of them were outside in the small area John had fenced off within their pasture so they could not stray too far and perchance have their babies in the field. Looking around I noticed a goat at the far end of the enclosure. Walking out to inspect, I saw that she stood near something motionless, white and flat on the hard, bare ground. I raced toward it. The doe seemed unconcerned, her back to what I saw was a kid, her kid. I scooped it up; it lay in my arms inert and limp. Noting that the mother was Tosca, a somewhat flighty but friendly animal, I turned and hurried to the house. There I turned on the oven, but the heat seemed too slow in coming. The kid was still alive but it didn't have long. Its mouth was cold; it made no effort to move. I remembered reading that if a kid is thoroughly chilled, one way to revive it is to immerse it in warm water up to its neck. It's best to put the kid in a plastic bag first so that its scent will not be washed off. If it is, the mother may not take it back. I searched the kitchen mentally; neither a large green trash bag nor a small vegetable bag would do. I'd just have to risk it. I drew warm water into the sink and slipped the little goat into it. When she had warmed up a bit I wrapped her in a towel and pulled a chair to the woodstove. I built up the fire that was already burning, then sat with her, still limp on my lap. It was too soon to give dextrose; the book cautions that a chilled kid should be warmed first before the sugar solution is injected.

The kid wasn't coming around. Some colostrum, I thought, might help, but I didn't feel I could safely put the still-inert body down and expect it to survive. Perhaps it would have, but at that instant, I felt I had to be there to pull it along by force of will. Just then there was a knock at the door. I turned to see who it was as John walked in. Could he help? Of course. He disappeared to the barn, where he set up a kidding pen and led Tosca into it. Then he stayed with the baby while I milked some colostrum from the doe. I gave it to the kid, but the instant miracle I hoped for didn't happen. It's thought that the mother goat stimulates her baby's nervous system to life by licking, which she does persistently and sometimes almost frantically at first. I tried rubbing the infant with a bath towel, quick, light strokes to simulate the doe. Slowly, almost imperceptibly, the baby began to come to life. At first she was too weak to do more than swallow a little milk. Then she began to lift her head on her own. After an hour or so she could remain upright without help. John stayed with me for a couple of hours, until the kid seemed to gain a purchase on life. After four hours, the kid could stand. Around nine o'clock I took her to her mother, and Tosca, in her unconcerned way, simply took up where she had left off earlier. She didn't quibble, didn't even seem surprised or concerned, but let the baby nurse and seemed sufficiently attentive.

I was grateful that she was one of the friendly goats. Motivated by greed as much as by curiosity, she responded to handfuls of grain from the first and had become used to me, and unafraid. Her baby, perhaps because she had already sampled her mother's milk in the first colostrum I gave her, suckled readily rather than rejecting her mother as she might have done at that point. But a bond, a kind of imprinting, formed in those early hours, and little Celeste later loved to follow me about, sometimes slipping under the electric fence to do so. I've often wondered if the success that evening of reuniting the mother and infant was made possible by their insouciant personalities. They're greedy and happy with a sunny immediacy. Perhaps this nature allowed them not to be adversely affected by their separation in the important first moments of life.

It was past nine when I finished in the goat barn. I went to bed and slept well for an hour or so. I checked the barn again around eleven. Tosca was settled in the hay, her baby nestled close beside her, her little mouth reassuringly warm. In the next pen Eleanor Rigby's baby cried intermittently. He had been suckling but now his mouth was cold. I felt myself gear up and draw on energy that must derive from maternal instinct. I milked the doe so that I could give her baby some milk with a bottle. But the nipples I ordered didn't fit the bottles. Instead I used a twelve-cc syringe, held the baby in a towel on my lap, and gave him milk. He drank readily and cried between swallows. I rigged up a heat lamp and held him close to it, still on my lap. At two o'clock I was still in the barn. At three he seemed somewhat weaker despite the nourishment. I returned to the house for a nap but was back in the barn at five.

It was sometime later that morning that I realized he would not survive. He simply wouldn't rally to the milk. Nothing else seemed wrong. There was no sign of respiratory problems that I could detect, although I had no way of listening to his lungs. Finally, after thirty-six hours of life, he died. Possibly his internal makeup was askew, preventing him from using the nourishment he took in. The doe looked forlorn and I felt sorry for her, but she appeared to adjust better than I, soon rejoining her companions readily. Perhaps she knew before I that something was wrong; perhaps the baby's insistent cry was wearing and told her more than I could guess.

A few days later when I was checking the barn I was appalled to notice one of the kids with a gob of bright yellow feces stuck beneath its tail, completely blocking the anus. I tried to pull it away but it was stuck fast, and the baby screamed in pain at my attempt. I went back to the house for a pail of warm water, a washcloth, and a pair of scissors. I took the squirming baby on my lap and worked with it, the mother standing in the pen, snorting and reaching over to pull at my jacket. After repeatedly softening the mass and snipping away the strands of mohair that held it, I was able to remove it—and I learned to check for this problem regularly. The first few hours, a baby goat's feces are black and tarry. Then they turn sticky and yellow, and it is at this point that they tend to adhere to the fleece at the base of the tail. After the first week of life, the kid's feces are tiny dark-brown pellets, miniature versions of an adult's.

At five days of age, the first-born kids were becoming active and were clearly ready for more expansive surroundings. John and I talked about what to do next. I needed a nursery area for the mothers and kids, away from the other adult goats, so John came over one early evening and quickly put up a partition, about four and a half feet high, down the center of the goat area and joining the large rectangular hay feeder. He made a door in the partition and cut an opening on the south wall of the barn so that the goats on both sides of the barn would have access to a pasture. We tagged the kids with ear tags and released them and their mothers into their new area. Then we watched the kids literally kick up their heels and jump and twist in the air, ecstatic at the enlarged space. They're fearless and friendly at this stage and curious about everything, particularly their peers, whom they're meeting for the first time. They dance up to inspect anything unfamiliar; they nibble at everything and try to drink from other moms, who send them scurrying with a gentle butt. Then suddenly, they'll settle down for a nap.

During that time, I loved to enter the goat barn in the middle of the night and see the moms and their babies, nestled together in the hay. I'd flick on the light and the mothers would look at me sleepily but would hardly stir while their babies continued to sleep. There was peace and serenity in the barn and a sense of priorities falling into place.

Kidding slowed with the post-midnight birth of Aida's boy on April 11. By that time there were seventeen births with fifteen kids surviving. During that time, Mimi grew from a rather small animal to a generous size, and with her growth it became obvious that she too was pregnant. She and Rita, both probably too small back in November to conceive, gave birth in May. Only two goats, Violetta and Butterfly, failed to have babies that year, possibly because they were both too small. I was pleased. It seemed that our goats had acclimated well in the short time they lived on the farm. Obviously the island suited them.

Male Kids & New Adults

OLD-TIMERS IN THE SHEEP- and goat-raising business maintain that female goats in a herd will coordinate their cycles of fertility so that breeding and ultimately birthing happen over a relatively short period of time. I've also heard that animals give birth during the phases of the moon, and I diligently watch the calendar to try to determine if this is true. I'm not certain of the veracity of either of these bits of folk wisdom, but now all our does, with very few exceptions, give birth within a matter of three weeks. And as if detecting my wishes, they obligingly, for the most part, come into labor during the day. The first year, though, birthing spread out over weeks, until the end of May. All that first spring, kidding focused my attention. My brain was so absorbed with learning animal husbandry and so alert to the needs of the goats that I had mental capacity for little else. I took a break from the piano lessons—the first lessons since childhood—that I loved; *New Yorker*s and the Sunday *New York Times* piled up untouched. Many nights, instead of bothering with a meal, I'd warm a mug of milk, add a bit of honey and nutmeg and go to bed with this soothing drink. Like piano lessons, I'd left off drinking milk years before; I could only think it was the maternal energy in the barn that awakened a forgotten taste.

The one—and considerable—concern that disappeared was my condominium in Santa Fe. New tenants were secured and I was again receiving an income and could pay back the loan I'd had to take out in order to stay on the island.

I had all but forgotten the new goats due in the summer, and the new barn to accommodate them. The last two kids of the year's crop were born in May, to Mimi and Rita; by early June we had seventeen healthy, fast-growing young goats, bringing our total to thirty-eight animals in a barn large enough to house twenty-five adults at most. Once kidding was over there was routine maintenance: I trimmed hooves, which took less time than before, and I scrambled to catch each now-hefty kid for inoculations. At first John helped me with the vaccines, holding the kids while I gave them their shots, but later I did it myself, graduating to solo as I had already with hoof trimming and worming. Each squirmy kid had to be cornered and pounced upon, sometimes missed but eventually caught and picked up, then secured, often between my knees to steady it while I jabbed a needle into the muscle of its hind leg.

The critical task of June was castration of the male kids. Our little males, lusty and sturdy, were already playing at butting heads and at mounting each other. A male kid can impregnate a female when he is as young as four months, but a female kid won't conceive until she is large enough in size, or sixty pounds according to the books, although some of mine have conceived when smaller. I wasn't concerned about the little females who were still small and delicate, but I did not want any out-of-season breeding of the adult does by the kids.

I called around for advice on castration despite knowing that most sheep raisers do not keep extra males. The few they have on hand, I learned, they castrate by banding, which is done with a special tool that slips a small, hard rubber castrating ring over the scrotum, close to the belly to cut off the blood supply. The trick is to make sure both testicles are down in the sac and that the band is not so close to the body that it blocks the urethra. The tissue of the scrotum slowly dies, becoming hard like a piece of dried leather and eventually dropping off. Another method is to clamp the scrotum with a Burdizzo clamp, a piece of equipment I felt reluctant to invest in without trying first.

I talked with Brenda in the veterinary office, who said their policy was to discourage banding. The dying tissue is an avenue for infection, she insisted, and it's painful over a much longer period of time than the preferred method of cutting the sac and pulling out the testicles. This causes pain for only a short time, she maintained. And I was in luck: One of the doctors was coming to the island on his semiannual trip to give shots to the horses kept by various islanders. He could stop by the farm and castrate our little males on that visit.

I'd do it their way, I decided. I had not heard any strong arguments for the other methods and no one on the island had a banding tool. We'd castrate ten of our twelve males, keeping intact Lucy's baby, Little Bill. For companionship, he'd have Leonora's kid, the first one born. Both were growing fast and had shot ahead of the others in size and weight. I was intrigued by the idea of developing our own breeding stock; I didn't want to wait for the day we might have registered females as well as registered billies. Who knows? If Little Bill turned out to be a great little buck, we might be able to use him or sell him as a fine sire.

"Let me know when the vet's coming," John said. "I'd like to see it, and I can hold the goats if you feel funny about it."

I was amused. Once when John was working on the house, I had asked him if he could dispose of a rotting deer head that the dogs had dragged onto the lawn. I'm squeamish about these things, although I'd been able to handle with aplomb the rib cages and hairy deer legs that also appeared from time to time. John bundled the head into a garbage bag and tossed it into the back of his pickup to take to the dump. I could understand that he now had reason to think the castration would make me shrink away; I knew otherwise.

When the vet comes to the island, he's scheduled nonstop from the time he arrives until the last ferryboat leaves. Word gets out that he's here; people who suddenly need to see him for an emergency call around to one horse owner or another until they locate him to put in a request for a few moments of his time.

Dr. Chris Koss, a young doctor who had recently joined the staff at the large-animal veterinary in Sturgeon Bay, was new to the island. He appeared hearty and kind and was immaculately dressed despite trudging around in barnyards and horse corrals. First off, he asked me to draw some hot water into his stainless steel pail, which already contained a blue disinfectant. We walked back to the barn where the ten little goats to be castrated were penned.

I grabbed a goat; it screamed piercingly. Dr. Koss squatted down and dunked his knife in the warm disinfectant solution and I knelt in front of him, holding the goat face-to-face with him. He grasped the furry little scrotum, and deftly inserted the knife horizontally until it ran through the upper part of the sac. The kid screamed again, but no more loudly than when I first picked him up. Dr. Koss drew the knife down, which opened the pouch like an unclasped pocketbook turned upside down. Then he pushed up the skin and found the testicles, two stringy membranes, pulled them out and tossed them aside. Last, he sprayed the area with a yellow antiseptic. It was all relatively bloodless. I released the kid into the barn and he scampered away, his little tail pressed firmly down as far as it would go.

"They'll be uncomfortable for a day or two," Dr. Koss said. "If they are noticeably down—if they walk in a funny way or stop eating—give them two cc's of penicillin twice a day for five days. It's best to do this when they're quite young, two or three weeks," he added. "The recovery is easier and their development is not slowed so much."

While I caught the next goat, Dr. Koss turned to John.

"Do you want to do one?"

Farmers usually take care of these operations themselves because it's simply not cost effective to have a vet come by for something that does not require any particular skill.

"Not me," John said quickly.

Then the vet looked at me. I needed to learn, but I preferred to watch him do another one before I tried. He castrated the second kid neatly, quickly. The worst part for me was the deafeningly shrill scream in my ear, but it was no worse than the screams of small kids at ear-tagging time.

"Want a try?" He handed me the knife. John held the goat. I don't remember which one it was. I took the soft, hairy little scrotum in my left hand and pulled it down taut. Then with my right hand grasping the knife, I tried to cut. I got as far as the point against the soft flesh but it wouldn't go in. I couldn't make it cut.

"Maybe you'd better do it. I'll watch."

Dr. Koss seemed to understand perfectly my hesitation. It was easy for me to imagine he felt the same way himself the first time he castrated an animal. He did a few more and I watched intently. I really should be able to do it myself, I thought; that's what a real goat rancher would do.

I took the knife again and this time succeeded in quickly piercing through the scrotum. I pushed the skin up and pulled out the slippery cords. Dr. Koss sprayed the area for me.

I didn't feel proud of myself, though; rather, I felt appalled. I have very little grasp of how a farm animal's nervous system compares with ours in conveying pain, but I have no reason to think they are less sensitive than we are, or than dogs and cats are, for example. It all seemed medieval. If these were house pets instead of farm animals, the owners wouldn't stand for it. They'd insist on anesthesia.

The kids survived the procedure. The youngest, month-old Bobby, scampered around immediately and seemed perfectly normal and unconcerned. The rest were subdued the first day but by the second had resumed playing goat tag and sparring with each other. Only two remained very quiet for a couple of days, hiding under the feeding trough when I came around. But when I wormed them five days later, every goat, without exception, seemed to have recovered.

The next year we had twelve males out of thirty-two kids. At a vet charge for castration of only four dollars an animal I had no intention of doing it myself. Peter agreed—he and Jeannine never stinted on veterinary care, and I always felt grateful for this. Later, though, at least five newly castrated kids looked off and ran slight fevers. So for five days, twice a day, I chased them around to catch them and give them penicillin shots. The following year I decided to try the banding method, which the shearers did for me at no charge. The kids screamed as usual but never appeared to be uncomfortable once released into the barn. And there was no sign of infection. The little scrotums eventually dried as hard as old leather and dropped off. I would find them occasionally, little hairy knots underfoot in the hay.

I looked to John for guidance in so many things that first year, even such simple questions as when to clean the barn. Horse stalls are cleaned out every day, I understood. But goat barns? Susie Waterman said she cleans hers once a year. When packed manure and hay are a good twelve or eighteen inches deep she comes in with a small tractor and scoop and cleans it out mechanically. We had no such equipment. The hay underfoot continued to pile up, and after a few months the door to the pasture was blocked open by the heavy layer of bedding. Winter came and in the cold, the ammonia smell from urine on packed bedding was not particularly noticeable, which was fortunate because if we had wanted to clean the barn at that point, deep snow would have prevented hauling the old bedding out. When we finally cleaned the barn the first time, the used hay had piled up for ten months.

John backed a large wagon to the door. Using manure forks, we took heavy loads of bedding out to the wagon. Underfoot were layers of hay pulled from the hay feeder and trampled by the goats, trodden with manure until it was a tightly woven carpet. John and I peeled back layer after layer of reeking, slimy brown hay woven densely and inextricably into a mass. The floor of the barn, the former machine shed, is cement, so the urine is never absorbed but remains to saturate the hay, causing it finally to exude fumes of ammonia. John and I worked from seven-thirty in the morning until four that afternoon, filling the wagon four times with hay laced heavily with manure, and hauling it to a neighbor who used it to mulch some new cedar trees. Standing on top of the load in the wagon, we forked it onto the ground, then went back to clean some more. As we worked, my forkfuls became smaller and lighter. If John had not been there I would surely have quit, leaving the barn half done. We could hardly breathe in the toxic air and later neither of us could smell anything but ammonia for several days. When we finished one side of the barn, I scraped the cement floor with a shovel, getting up as much as I could of the last layer of slime mixed with barn lime. Then I sprinkled fresh barn lime, or calcium carbonate, on the floor to help absorb odors, and on top put down a fresh layer of hay. When we finished, I could do little more than feed the goats, bathe, and go to bed. John must have been just as tired because later Patty insisted he never clean the barn again.

What was apparent was that it had to be cleaned more frequently. I was concerned about the goats. If the ammonia fumes were bad for us, what about the animals, many of whom stood only some eighteen inches above the bedding? After three months I borrowed John's wagon and cleaned the barn again, but that still took me a good eight hours. Next I tried a six-week interval and tried to pace myself by stopping for frequent breaks. It was a vast improvement, but still not good enough. Now I clean the barn that has the cement floor every two weeks and sometimes more frequently. The job takes from three to four hours, or more if I let an additional week elapse. It's no longer a grueling job, because in that short length of time, the hay and manure do not weave themselves into a dense carpet that literally has to be torn apart.

My sister-in-law Jeannine's baby was due in August. Jeannine, Peter and Anne came up for occasional weekends that summer but did not spend any length of time on the farm. For my birthday in July Peter paid for me to go to a workshop on raising goats to be held in August in Hastings,

Michigan, near Kalamazoo. There would be an all-day shearing class, which was the main attraction for us, still detemined as we were to do our own shearing. He and Jeannine would go, too, he said optimistically. Later, he thought better of it; perhaps Jeannine would stay at home but he and I would certainly go, he insisted. But he never filled out the registration form for classes at the workshop. As July wore on, it seemed apparent that he would stay with Jeannine, which was only right.

In the meantime, we were slowly preparing for more goats by clearing additional land of brush so that it eventually could be turned into a pasture. Peter had bought half interest in a Brush Hog, a piece of machinery that is pulled behind a tractor to cut all manner of bushes and small trees. My niece Susan used it, pulling it behind the little Ford tractor, a gray and red antique. When I wanted to learn, Peter said he'd rather I didn't. I mentioned this to John.

"Susie, you can do anything you want to do. I'll show you."

John's a careful, patient teacher. The tractor is stiff and temperamental and doesn't always start. But he showed me how to run it and how to coax it along when it balked, then let me practice in the field east of the house. Next he hooked up the Brush Hog and had me pull it without actually cutting anything. When I had mastered it, I engaged the blades and dragged them over the sprawling, tough juniper brush and the chokecherries that grow like weeds. The Brush Hog hesitates when pulled over small bushes and trees, vibrating and grinding for a few moments, and in those seconds I wonder if it will break down. Will the blades defeat the growth or become entangled in it? It's short-lived suspense because with a final grunt and grind, it moves free of the demolished plant. I spent many hours for many days on the tractor. Later, I cut the purple spotted knapweed that colors the fields in August, but is so tough and stringy that most animals won't touch it. The field would become a pasture in a year or two, and we wanted to cut the knapweed before it bloomed and went to seed, sowing another generation of plants.

The barn was crowded. There were twenty does and a billy, all nearing their mature size. Some of the smaller does, Mimi in particular, had had a sudden growth spurt and were turning out to be generous-sized animals, broad and large in stature, and some of the seventeen kids had grown nearly as large as their mothers. Lucy's son, Little Bill, and Leonora's son looked absurd when they scrunched down on their knees to take a drink from their mothers. They were the consummate mama's boys. The mothers had little patience for this and usually walked away abruptly as soon as their huge offspring had taken a short swallow. At times the barn literally bulged with thirty-eight nearly full-size animals.

There was plenty of rain as well as warm sun that summer and the grass grew well. It was a pleasure to see the growing kids, their fluffy fleece white against the emerald green of their pasture. From my vantage point on the tractor as I worked nearby, I'd exult, luxuriating in a feeling of peace mixed with pride in our growing herd.

Time was growing near for the completion of the arrangements Peter had made with Susie Waterman in March to acquire new goats for us. She was to deliver the billy he'd chosen, and as many as twenty additional commercial-grade does that she would travel to Michigan to select. As the day approached for the additions to our herd to arrive, I began to be concerned about overuse of our pasture, as well as overcrowding. Our little enclosure was only five acres and despite the more-than-adequate moisture and sun, the grass could not recover once the goats had eaten it down, at least not without resting. Soon we'd be feeding hay again, although the goats much preferred to nibble at the now close-cropped blades. I didn't know when Peter would enlarge the pasture but he'd said we'd build a new barn adjacent to the old one. He had plans for it, sketches he'd made, and John would do the building. But in the meantime, Susie would come soon with the new animals, increasing our number to as many as fifty-nine. It seemed unimaginable. Couldn't we get them later? I felt I was pleading, but Peter—five hundred miles away in Indiana—remained unconcerned.

Mid-week at the end of July, Susie drove up with the goats. She had been able to find only twelve does that were suitable and had crowded them and the billy into a specially outfitted station wagon for the trip from Michigan. They were all yearlings, not yet full grown, and appeared a bit forlorn after their days on the road from Kalamazoo to Madison and then up to the island. Susie backed the station wagon up to the barn. As each animal emerged from the vehicle, we inoculated it and gave a precautionary shot of Oxytetracycline in case the stress of the trip had weakened their immune systems. Then we went out into the pasture to share in a little goat watching.

The new does bunched together, cohesion produced by physical proximity over the last days and by the looming presence of our original animals. The new ones moved as a group, edging farther into the pasture, nibbling grass nervously, it appeared to me. Occasionally one would butt another; there was much defensive snorting. The new billy, whom I named Jose, was not so tentative. Entranced, it would seem, by twenty new girls plus his traveling companions, he rushed around inspecting does. Violetta, one of the two Texas does who had not had kids, seemed particularly tantalizing to him, and he repeatedly waved his nose in the air and curled his lip at her scent, then poked at her, chortled, and pawed the ground. I wondered if we'd see a December kid.

The most expressive animal by far, though, was Bill, the original billy. Astonishment was the primary emotion he conveyed. He seemed simply not to be able to fathom a competitor, but rather than focus on this masculine intruder, he was distracted by the charms the new girls held for him. He ran around wildly, without direction, snorting his aggression and posturing one moment, then flapping his tongue and chortling at the females the next. He seemed both incensed and intrigued, hurt and delighted. I had betrayed him, insulted his very masculinity by introducing a rival, but at the same time had served him up a delicious assortment of new young females unencumbered by offspring.

The two billies circled the does, remaining on opposite sides of the pasture, never coming too near each other. Like male dogs, stiff-legged, they took stock from a safe distance. If the does had not distracted them, they surely would have sparred, rising on hind legs, dancing with heads tilted to the side, then clashing horns and heads. However, it seemed they had only half or quarter of a mind to do so, their attention pulled more delightfully toward the novel females.

Susie and I sat on the grass watching and talking goats until evening began to fall. Just as the light faded in earnest, Colin, a Canadian who lived on the island and was fast becoming a friend, came by to see the new arrivals and fête us with a celebratory bottle of wine. In the peaceful dark, we three listened to the night sounds of crickets and frogs; later, Susie and

Colin kept me company around the kitchen table while I cooked dinner. Colin's enthusiasm and wit, now trained on what Susie and I loved best, was flattering and drew us out. Conversation tripped from one subject to the next until it was well past midnight. By that time Colin even talked of getting some Angora goats himself.

Shearing School

A LIGHT WIND stirred the dust on the broad parking area of the Barry County Fairgrounds at Hastings, Michigan. At eight o'clock the sun was already warm; it would be hot again. Well before dawn, at the large cement-block bathroom at the fairgrounds, women chatted over the washbasins about their livestock and the workshops they'd attended the day before. I felt almost like a pro. One kidding behind me, I could share this experience and others with new Angora goat raisers. Dressed in jeans, hardly a touch of makeup, we were as excited as first-time campers. We didn't hesitate to initiate conversations with each other. All it took was a simple "What do you raise?"

It was the second full day of Fiberfest, the annual fiber show and collection of workshops that draws people who raise goats, llamas, sheep, Angora rabbits, and alpacas from all over the Midwest. The main reason I was there was the eight-hour Angora goat shearing school given each year by Willie and Joe, a father and son team of shearers originally from Texas, roots in Mexico, who would teach us novices how to shear our goats.

John didn't want to shear again; he'd made that clear. It

was hard on his back, and besides, he pointed out, at his carpenter's hourly rate it was needlessly expensive for Peter. The solution, as Peter saw it, was for me to take the day-long shearing class offered in Michigan in August.

I enlisted my twenty-year-old niece, Susan, to come to the island to care for the goats, and set off in the black Dodge van on the ten-hour drive to the conference. I'd sleep in the back of the van, parked with RVs, pickups and other vehicles on the bare, dusty lot at the county fairgrounds.

The shearing class convened in a long shed, open on all sides. There were about a dozen of us, most in pairs, all of us sporting shiny new shearing machines. Mine, I observed, wasn't quite so new; at least it had enough goat hair and grime in the crevices to prove it was a veteran of a couple of shearings. The students looked as newly minted as their equipment, and I wondered how many had actually tried shearing. A few, no doubt. A couple from California was there. The day before at the workshop given by a veterinarian and goat raiser, the woman had raised her hand and asked how to prevent the goats from urinating in the barn. "Very hard to do," replied the vet politely. I was certain this was a first-time event for them.

A stocky, swarthy man introduced himself as Willie in the slight lilt of the Hispanic Southwest. Helpers included his son Joe and a few other experienced goat shearers who wore cowboy boots, Western shirts, and faded jeans. A pen next to us held a group of raggedy-looking goats; obviously we novices would not learn on anyone's prize animals.

Willie demonstrated on a couple of animals. A man of fifty-five or sixty, he was handsome in a rough-hewn way, with a kind and ever-so-slightly flirtatious light in his eyes; Anthony Quinn as Zorba the Greek, I thought. He'd shear

Western style, he explained. Joe, a man in his early thirties, brought in a goat, picked a few pieces of hay from its long, scraggly fleece, and positioned it in front of Willie on two plywood panels. With a quick, fluid motion, Willie brought the animal down on the plywood and turned it on its back in front of him, one knee helping to pin it down. The goat showed no inclination to fidget, probably sensing Willie's expertise and the futility of any attempt to move. First you open up the belly, Willie explained, shearing around the udder, a finger on each teat. "If you shear the teat off a rancher's prize doe, he's not likely to ask you back," he cautioned. The goat lay there as if in a trance. With an easy, broad sweep he ran the shears up the belly to the breastbone and beyond along the inside of the forelegs, the hair falling away cleanly. Next you do the inside of the legs, he continued. With a deft movement of his hand on the animal, he caused the goat to straighten her back leg, which he sheared clean on the inside. Then the other hind leg. It all looked so easy. Next you hog-tie the animal, he instructed, bunching all four legs together at the ankle and wrapping a bit of rawhide around them.

He sheared around the face and throat. It seemed he knew the animal so well, sensed its contours and the loose folds of its skin, that he hardly looked at it. The machine glided smoothly over the doe; the hair fell away leaving bare pink skin. There was never a second stroke. When the face and neck were hairless, Willie rolled the goat into a shaggy lump at his feet and sheared from tail to neck with a few smooth passes, leaving the animal in a large smooth ball like a mound of well-kneaded bread dough. Willie stood up, stretched and let Joe release the goat from its bondage by a quick jerk of the rawhide. He led it back to the pen and returned with another animal for demonstration—another swift, flawless job.

The second goat, shorn a glistening silvery pink, scampered back to its pen, a signal for us to regroup in ones and twos at stations indicated by plywood sheets. An experienced shearer stood by to help each student or couple of students. Joe stood by my station and helped me set up my shearing machine. I laid out combs and cutters, screwdriver, and Three-in-One Oil.

"That oil's too light," Joe said first off. "Use this."

He handed me a small oilcan with a long spout. If you use a lightweight oil, he explained, the machine will heat up too much and be hard on the goats. He watched me insert the comb and cutter and adjust the tension knob.

"You sheared before?" he asked, not unkindly.

"Well, I tried," I answered. "A couple of times, with the help of a friend. Twenty-one goats each time. We didn't do a very good job," I added.

He smiled. "Well, go get your goat and we'll see. I'll be here with you to help out."

The goats waiting for us must have been old and experienced, not skittish like ours at shearing. Grabbing one by the horns and dragging her back was no problem. So far so good, I thought. Getting her down wasn't so easy, but I managed to trip her and roll her into position. It had been five months since spring shearing and I didn't feel all that calm. I remembered what Noel had said: "You have to shear a thousand animals before you feel comfortable at it." Joe squatted at the goat's head and helped hold her. I started tentatively around the udder area, a finger on each teat. Then I started to sweep up the belly but at about halfway began to lift the shears, leaving more hair than I should. "Keep them flat on the animal," Joe said. I tried with the next sweep. It seemed better, but still the cutting head rose slightly midstream. "Don't worry about cutting the goat," Joe said. Even though I accepted intellectually what I should do, my hand drew back, pulled up by caution.

Next I tackled the head, Joe showing me how to tilt the head away from me to get a better surface along the neck and jaw. I sheared tentatively, carefully around the eyes and ears, and cleaned up around the horns as best I could.

"Now hog-tie her," Joe instructed, handing me a length of rawhide. I bundled the legs together and tried wrapping them but I wasn't sure what to do about securing the package. I looked to Joe.

"Just wrap it the way you'd do a feed sack," he said.

"Umm. I'm not sure how."

Joe laughed and showed me. "Now you do it."

I tried awkwardly a couple of times, then got it right. Joe found this vastly amusing and joked throughout the session about my feeble knot-tying technique. Shearing the bundle of goat against the grain was easy if you didn't think too much or try to analyze which end was which. Joe seemed satisfied with my progress, though, and instructed me to fetch a second goat. That morning I sheared three before we broke for lunch.

The group that convened after the meal had whittled down to five students.

The California couple did not return, nor did some of the others; no doubt they'd begun their search for professional shearers. Joe helped me again. Concentration propelled the clock; time flew. At four-thirty, when the group began to break up, I'd sheared four more animals. Willie had been watching Joe and me much of the afternoon.

"If you want to stick around and do another, that's OK with me," he offered.

"Great," I said. "I'd like that." After all, I thought, I've got a long way to go to a thousand goats. I'd like all the supervision I can get.

I sheared my last goat; it didn't seem noticeably easier than the first. I'd still get to places where I wished and prayed I would just get out of it with the goat unscathed. But they all survived and there weren't even any cuts. Joe and Willie seemed pleased that I'd stayed until the end.

"We're having a barbecue for our friends tomorrow night," Willie said. "How about coming along? Just bring a six-pack," he added.

The next evening I put on the last clean clothes I had with me and strolled over to where Willie and Joe were camped, in the parking lot not far from where I'd stationed the van. The air was still warm, but not with the afternoon's baking heat. Over the dusty parking area a breeze wafted the pungent, smoky aroma of mesquite from the barbecue grill. A knot of cowboys—men in well-worn Western boots and jeans and equally well-worn Texas drawls—clustered around long tables set up beside pickups. Some wore cowboy hats. There were women, too: middle-aged women in white pants, younger ones, long-legged in short shorts. A few small children played by the tables; dogs slept off to the side. In the midst of the groups, seated at one of the picnic benches was Willie, Zorba-style, surrounded by young women. He looked up, smiling, and waved me over.

"Glad you came," he said. "Grab a plate and something to drink. There's tequila and Dos Equis. We'll have goat fajitas ready soon. Ever eaten goat?"

I thanked him and put the beer and salsa and chips I'd brought on the table. There were bowls of guacamole and salsa, bean salad, macaroni salad, bags of chips and white flour tortillas, soft drinks and pitchers of lemonade as well as an assortment of brands of beer. Again, it was easy to enter into conversation or at least to listen and feel a connection because most of the people were talking Angora goats. Many, I surmised, were from the San Angelo area of Texas. I knew that some of the judges for the main Angora goat events were from the big goat breeding families of the Edwards Plateau, and many of the people at Willie's barbecue seemed to have Texas roots—or at least were Texans for a night. I met other people, too, such as those from the surrounding area. They were people who love goats and had been coming to Fiberfest from the time it started a few years before. One woman and her husband raised goats while he taught school and she studied nursing. They were like many others I met who loved the animals and had to work hard in order to afford to keep them. We exchanged addresses, and later a note or two, but eventually lost each other.

Joe handed me a plate with a white flour tortilla and some strips of meat hot off the grill. "Goat fajitas," he announced and pointed to a bowl of chopped green chiles and some red chile sauce to eat with them. The meat seemed overcooked to me, a bit tasteless and dry, and the sauce biting compared with the relatively bland fare in Wisconsin. But I ate the fajita and tried not to think of Ariadne and Carmen and Lucy. Perhaps they were Spanish meat goats, I thought. Angora goats seem too scrawny to be a commercial food source.

Plate in hand, I made my way toward my host.

"*Habla usted español?*" I asked, certain that he did speak Spanish.

"*Sí, claro,*" he answered in the affirmative. "*Sientate,*" he said, inviting me to sit down and join him.

We continued speaking Spanish, mine rusty, his smooth

and flowing, until the words began to fail me. The party felt like a bit of New Mexico or Texas right there in Michigan. I looked around but didn't see any of the other people who attended the shearing school the day before, but most of the people who assisted Willie were there. Susan Drummond, who organized Fiberfest, wandered over, an old friend of Willie's and Joe's. I left the gathering that evening feeling nostalgic for the hospitality of the Southwest, and excited by listening to the adventures of goat ranchers who ran huge herds on the Texas range, experiences that those of us with our tiny pampered flocks in the north could only imagine.

The few days were packed with information by way of seminars on basic goat management, on veterinary aspects of goat raising, and on the marketing of fiber products. But also there was a constant exchange of experiences with other novice goat raisers which charged my enthusiasm to a level of those first exciting days of our project. I spent the last afternoon looking at goats, llamas, and sheep—penned, combed and awaiting judging—and browsing through the great hall where booths were set up selling everything from books and dyed yarn to fencing supplies and shepherd's crooks.

I left before light for the ten-hour drive to the ferry dock. Brimming with information and new ideas, I could hardly wait to get home to put them to use. I'd call the co-op and have some lengthy discussions about a good mixture of vitamins, minerals, and added protein for our feed; I'd start Antonia, the new doe from Michigan whose knees buckled, on supplements. I'd already called my niece to give her instructions for injecting Antonia with selenium, which was suggested by one of the vets at the workshops. Susan had never injected a goat before—or anything else as far as I knew—but that wouldn't daunt her. As I drove home and reviewed the seminars, the exchange of experiences, and the tips I learned as well as imparted, I realized I'd graduated from rank beginner, as wobbly and wet as a new kid, to successful veteran of one complete year with the animals.

But driving along I also had time to reflect on the real purpose of the goat conference: the shearing school. I was to come away from it a self-confident shearer, capable of handling our herd singlehandedly. I'd done eight goats that day under the guidance of Joe and Willie, who advised me when I was uncertain how to proceed, but it was clear: the eighth goat hadn't sheared too much more easily than the first. And I had no way of inspecting the sheared fleeces to see if they looked better than the uneven products of John's and my efforts. Our fleece had recently been rejected by the woolen mill where I tried to sell it. "You'd be well-advised to get a professional shearer," the mill director had written to me. With the additions to our herd, including two old wethers Peter had bought from Ingrid, I'd be shearing fifty-three goats in September, not twenty-one. As I drove along I resolved to search out a professional goat shearer and convince my brother to pay whatever necessary to get him or her to come to the island.

Making Strides

OUR SHEARERS, Larry and Frank, now feel like old friends who visit twice a year and just happen to do two days' worth of much-needed work. The September after the shearing school I engaged them to come up from the Madison area to shear our goats. They set to work in the barn, with Larry, a man nearing sixty, shirtless and burly, bending over each goat that nineteen-year-old Frank caught. They were gentle with the animals and, I think, tender-hearted toward them, and they graciously endured an audience of my friends who wanted to see the shearing. John and Patty came by the first night and John stod there beaming as he watched. I understood his smile: a combination of admiration for Larry's skill and relief at not having to shear the goats himself. I felt the same. That shearing and at later shearings I always worked alongside Larry and Frank, sweeping up, separating out the stained fleece, bagging the hair, and in the spring inoculating the animals. By not actually shearing myself, I had time to evaluate the physical condition of each goat as her hair came off and to exult inwardly over the beauty of a particularly lustrous or heavy fleece. The shearing took two days, the afternoon and all evening of the first and into the afternoon of the second. I put the shearers up in the guest apartment and cooked the best meals I could in order to keep Larry and Frank willing to return. They ate prodigious amounts, Larry—a short-order cook when he's not shearing—complimenting me with the assurance of someone who knows his way around a kitchen, and Frank giving me a quick nod, but shyly not meeting my eyes when I offered him more. We would sit around the kitchen table until late at night, talking goats and sheep. One time they brought a Parcheesi game with them and knocked quite late on the kitchen door just as I was finishing the dishes. They had set a place for me at the board game, they said. I joined them, touched by the invitation, but nevertheless longing for sleep.

I'm now careful to arrange the shearing months before the actual date because to clip our goats, Larry and Frank need to work three days into their schedule, allowing for six or seven hours' travel each way in Larry's aging pickup. I'm grateful they make the trip for it hardly seems worth their effort, particularly for the two dollars per goat Larry charges and the six dollars an hour for Frank's time and the extra we pay them for the trip. But I think they consider it a bit of a getaway and enjoy staying at the farm.

By the time Larry and Frank came to shear that summer, John had started work on the addition to the goat barn, which ultimately tripled the original space. I was his carpenter's assistant, partly because the experience intrigued me and I like working with John, but also because I needed the money. Once the footings and frame were in place, John set me up on the lawn with a table saw so that I could cut battens, which I nailed vertically to cover the seams between the boards that made the barn. I learned a new vocabulary along with new skills: not only soffits and battens and names of new tools, but also that I was ripping boards when I sawed them and using a bubble stick when I checked a surface with a level. Much of the time in my association with John I lacked the words to discuss mechanical or carpentry problems, and he was always amused to see me use my hands to describe what my words could not.

We watched the sugar maples change from green to red to gold against a deep-blue sky and from the roof where we stapled shingles, we paused to see huge flocks of geese fly south. We moved the goats into their new quarters October 18. A few days later Nora, who raises sheep and a few goats and trains herding dogs on the peninsula, delivered two bred does and their doe kids from the previous spring. Our herd then numbered fifty-seven. Later, in November, when the weather turned cold, John and I gathered beach stone to make a chimney on the west end of the machine shed, the original goat barn, which we had recently insulated and lined with wallboard for what was to be an office. John erected scaffolding and we both worked, he putting on stones and mortar and I scraping out the excess mortar, defining the stones as if they were sculpture. It was then that the winds whipping us on the scaffolding sent me to Wal-Mart in Sturgeon Bay for insulated coveralls, which became my outdoor uniform, night and day, much of the year.

Colin died suddenly that autumn. I had come to look forward to his enthusiastic telephone calls to inquire about the goats and what I was doing.

He would drop by with clippings from the *Wall Street Journal*; I would reciprocate with *New Yorker* articles. Occasionally he invited me to see a video from his collection of foreign films. One weekend much earlier in the year, when Peter was on the island, we had the first of several gala and memorable dinners, full of friendly sparring and Colin's witty banter. Just a week before his death he had helped me trim hooves by holding the goats for me. Two of his grown sons, one from Canada and the other from Illinois, stayed for a few days after the funeral, and I invited them to dinner along with my island friends Dan and Julian, who knew them. It was an evening Colin would have enjoyed; we sat around the dining room table until nearly three in the morning telling Colin stories. For a long time after his death the island seemed gray and bleak despite brilliant weather. At Peter's suggestion I took on the chore of feeding Colin's four Icelandic horses, which remained for a year at his farm. Twice a day I would drive over to the place and feel its emptiness and desolation and I wondered if the horses were puzzled at their abandonment.

For our second round of breeding I assigned the new billy, José, to the new does, adding Violetta and Butterfly, two of our original group who had not given birth the previous spring, and Gilda, whose kid had died at birth. I put Bill with the rest of the Texas does. To add some order to the chaos of breeding and to give us a better estimate of when to expect kids, I bought two breeding harnesses, biblike affairs that hold a chunk of crayon and are worn by the males. The colored wax rubs off on the female's rump when the male mounts her. Bill was blue; José red. The boys got to work quickly. Several times a day I'd note which does had been marked, but the smudges were ambiguous. I'd see a red one on a shoulder or a side, or a very pale blue mark on the hindquarters. Did this indicate merely a feeble attempt or an altogether futile one? Perhaps the female wasn't ready. For two weeks I watched and jotted down the numbers of does with fresh smudges, then calculated when they should give birth. It all seemed so easy.

By that time my attention was diverted to other things. Our hay was molding, and becoming dusty from having been baled too wet. It was not fit for the goats and had to be hauled out of the barn. More serious, the automatic waterer again gave me a mild shock when I scooped out dirty water. I watched and saw goats attempt to drink, then pull away abruptly. I had to find the source of the problem, which turned out to be a fault in the way the fence was grounded. And the fence wires kept wearing out and breaking. Between fiddling with the fence, hauling water and hauling and restacking hay, the antics of mating became routine.

On November 5 the weather turned cold; wind blasting from the north sent the first snow hurtling across the bare fields. When I checked the goats I noticed Bill outside the barn despite the snow. He was eyeing two of José's girls. On the other side of the fence, Musetta and Salome flirted with him. Both had been duly marked at least two weeks before by José, but apparently had not conceived or they would not be cycling again. An hour later I stopped by the barn. Bill was neither in the barn nor out in the pasture.

Then I noticed the fence was down. He had burst through the Polywire, breaking some of it and popping other strands from their hooks to get to the females of choice. I unplugged the fence charger and went to work.

Polywire fences are easy to repair, which is fortunate because they break down just as easily. The twists of plastic and thin wire snap with a minimum of wear and some pressure against them. The fence held well for the first year but by the second it was giving way, especially to a large determined buck in the throes of lust. It seems to be the rule of farming that a fence will break down in the worst weather. When the wind blows furiously and the snow falls thick and fast, you can count on a Polywire fence needing immediate repair out on the far side of the pasture. This time was relatively easy. Once I had it working again, I grabbed Bill by the horns and, pushing and dragging, coaxed him to his area.

The snow did not let up; the wind howled furiously around the barns. I checked the goats after lunch. Bill's girls were alone, their consort nowhere in sight. Outside I could see he had broken through the fence again. Again I fixed it, taking fresh lengths of Polywire to tie onto the existing wire. Then I went in search of Bill on José's side. He knew what my presence meant and he didn't like it. I was interrupting his love life, inserting myself between him and the object of his lust. He reared up on his hind legs, lowered his head slightly to the side, and danced like a boxer in the ring threatening a parry. Angry, he was challenging me, vying for dominance. I grabbed him behind the horns, trying to stay clear of them, and wrestled him back to his side of the barn. I felt shaken; I was breathless and angry, too. Salome appeared to be the doe most interested in Bill at that point, so I led her to his pen. Anything to keep him content on his side of the barn. But Salome wasn't enough to satisfy him that afternoon. A third time I fixed the fence, my fingers numb with cold, the wind whipping against me, tearing at the loose ends of wire as I put the fence back together. I gave in and decided to take the easy route. I led José into Bill's pen and let Bill remain with José's girls. In the short term it would keep the fence intact in brutal weather. In the long run, it spelled difficulty: I would never know with certainty which goat fathered which kid and therefore would not be able to breed José or Bill with the new generation without risking inbreeding.

Goats prefer their accustomed surroundings, I've learned. The next day Bill and José were back in their original pens, having gone through the fence one more time. This musical chairs in the barn repeated itself several more times, leaving many does with both red and blue smudges and ensuring that I'd never be absolutely certain of the paternity of the new crop of kids.

One day when Bill seemed satisfied on his side of the barn, I noticed Delilah standing at the fence shaking her tail at José on the other side. Bill spied her and raced up, flapping his tongue and chortling. But she showed no interest in him. She moved away and continued flirting with José. So goats have preferences, too, I observed. I felt the least I could do was further the romance, so I removed Delilah to José's pen and let the courtship proceed unimpeded.

Five months later when the kids were born, the date of the crayon markings had little to do with the order of births. The only notable exceptions were two or three does that Bill marked first off, which were among the first to give birth. With all the switching back and forth between pens that fall, I could only guess which billy had sired whom. But I suspect the red smudges left by the young José in those first days of breeding had more to do with his enthusiasm than with the females' readiness.

November 7 the temperature dropped to fifteen degrees; more snow blew down from Canada, swept across Lake Superior and blanketed the villages of the peninsula until they looked like Christmas-card scenes. We had our first casualty among the mature goats then. Eleanor Rigby slipped on the ice near the electrified fence and became entangled in the Polywire, the strands wrapped around her horns. Outside I had first heard measured bleats, so precisely timed they sounded mechanical. I found her stretched out on an icy bank, baaing with every pulse of the electric current. Tom, Peter's closest boyhood friend, was visiting, and he helped me set up a kidding pen and lift her into it, where she quickly declined. Her brain had swollen with the shock, the vet said, and despite treatment with Banamine and steroids, vitamins and penicillin, she never recovered from a coma. After three days I dug a grave and John put her down, with a shot from a twenty-two-caliber rifle.

That fall I deliberated about what to do with the fleece. It was clear to me that there would be no immediate return on Peter's financial investment and my investment of labor. Peter's design for washing mohair over a wood fire existed only on paper, and the fleece I sent to the woolen mill to be made into roving for spinners sold hardly at all. I considered stockpiling our mohair until the day came that we were ready to use it, but I did not want to risk its becoming nesting material for mice and other creatures. It seemed better to ship it off to Texas for sale on consignment, then apply for the mohair subsidy. Susie Waterman's farm was the Wisconsin pick-up point for an Angora goat ranchers cooperative, based in Michigan and established to transport fleece to Texas and oversee its sale. I sorted our mohair according to the age and sex of the animals and bagged it in burlap feed sacks, labeled them according to specifications, and sewed the bags closed. Then I loaded them in the van and drove them to the UPS collection site at a general store in Bailey's Harbor on the peninsula. From there they would go to Susie's farm and then to Texas. In January I received a check for $396 for the sale of 372 pounds of mohair from our fall clip. It came to twenty dollars more than the cost of the fall shearing without figuring in the shearers' meals. The following April a thousand-dollar subsidy government check arrived; it was not much for a year's work, but we considered it merely an interim bonus.

The real income, Peter and I were certain, would come eventually from processing the raw mohair and having it made into consumer products.

Early one morning a month after Eleanor Rigby died I found Carmen stretched out in the hay, motionless. I dragged her to a kidding pen, took her temperature, which was subnormal, arranged a heating lamp over her and called the vet. "You've got to get her on her feet and get her temperature up," he said. I lugged and pulled and prodded, but she refused to stir. It was then I remembered learning that when the sheepherding islanders of Greece are confronted with a sick animal that won't get up, they give it brandy. We had no brandy at the house, not even the Remy Martin cognac Peter prefers, and the liquor store would not open until afternoon. I called our neighbors; perhaps Ray would have a bit of brandy on hand. His son-in-law Jeff answered, then searched the cupboard and reported a bottle of blackberry brandy. I tore over, small jar in hand, and collected a bit, much to the amusement of Jeff, who later teased me about early morning tippling. With a teaspoon I dribbled a few drops, and then a mere trickle into Carmen's mouth. The alcohol must have warmed her and the sugar provided instant energy because she raised her head immediately, licked her lips, and struggled for more. She slurped some from the spoon and allowed me to pull her to her feet. She seemed somehow dazed; there was a vagueness about her movements so unlike her normal assertiveness. I waved my hand in front of her face; she did not react by so much as a flicker of an eyelid. Unlikely as it seemed, she appeared to be blind. Yes, the vet agreed, blindness is a symptom of polioencephalomacia, which is not related in the least to polio in humans. It's caused by a microorganism that is usually found in a bit of moldy feed. It causes a sudden depletion of thiamine and consequent swelling of the brain. I began giving her Banamine to reduce the swelling and high-potency vitamin B shots for their thiamine.

Carmen regained her eyesight in a week or two, and after a setback of a high temperature that I treated with penicillin, she seemed almost well. She never recovered her old deliberate personality, however, remaining slightly addled and tentative. I kept her away from the others for much of the month.

But she had company to lift her spirits, for the two bred does we bought from Nora gave birth shortly before Christmas to female kids whom I named Noel and Navidad. I kept their kidding pens with Carmen's in the designated office area.

With Carmen out of danger and the only early arrivals well launched, I could enjoy Christmas without having to hover mentally over the barn. Glistening snow on starry nights, pine boughs and decorated Christmas trees glimpsed through glowing windows, sweet smells of holiday baking—cinnamon and clove and vanilla. It's a time of visiting, of relatives coming from across the water, of church pageants and Christmas carols. Peter and Jeannine always stayed in Indiana, visiting my sister-in-law's large family, but I never lacked a place to join festivities at Christmastime. On the island, Christmas seems more magical than elsewhere because of the lack of commercialism. For the most part, Christmas shopping must be done off-island and generally dispatched in good time lest severe weather prevent boats from crossing. There is no last-minute scurrying and jostling for one more present on Washington Island; at five o'clock on Christmas Eve Main Street will be deserted, the few stores and businesses having closed early so that all can attend Christmas pageants at the two churches.

Snowstorm

SNOW FILLS MY mind, blows into the crevices, balloons into large spaces, spills over the edges and pushes out everything else as it literally takes up my waking and sleeping hours. At first it was a blizzard. Strong winds from the northeast sent icy chips hurtling horizontally across winter fields. I gasped for breath as I went out to the goat barn and had to swallow to suppress a feeling of being suffocated. I tried breathing through my mouth; even that was difficult. I felt that the dense and driven particles that stung my face left no air for me. The snow and wind have become entities with personalities that range free and wild across the land. They're not playful as they sometimes are; instead they're crazed and sinister. After feeding the goats, whose backs were frosted with snow that blew into the barn through cracks in the doors and holes for ventilation, I settled down to watch and wait. Through the window I could see waves of snow, many feet deep, blowing across the landscape.

By thirty hours the wind had lessened somewhat and the flakes were larger. No longer a blizzard, it was simply a heavy snowstorm, but still pervasive. Mike, the young man who plows the drive, arrived with a companion. The plow moved masses of snow until it was mounded six and seven feet deep in places. Mike skillfully extricated the van and cleared a large patch near the house, framed by piles of snow. I threw on a coat and ran out to ask him to clear in front of the goat barn, the easier for bringing hay to the animals. It was still snowing, but at least I could now use the van to mail letters at the post office and buy groceries. While I drove the van, I looked down from my high vantage point and saw a mist of fine flakes swirling in my path, like carbon dioxide wafting above dry ice. It was the snow and it was dancing a furious, manic dance.

Next morning, large white flakes continue, but now they're propelled gently from the southeast. The driveway has filled in again. The snowdrifts lining the drive, last night's jagged mountains, now mound softly, ice–cream–like, edges smoothed and seams filled in by a new glistening layer. I clamber through hip-high drifted snow to the low stone fence at the edge of the lawn. I'm wading through thick marshmallow fluff while my dog bounds ahead, then waits on the other side of the field, wondering why I've suddenly become so slow. I go back to the shed for snowshoes. Even they sink ten inches or more. I labor across two fields, then turn back. I've had flu and still tire easily.

Today I start feeding John's horses and tomorrow, Glen and Kim's dogs and cat, and I worry about clearing the driveway, about getting out and possibly skidding the van off the road, miring it in snow. I do not want to refuse to help these friends. John never fails to come when I need assistance, whether it's putting up a barrier that the billy goat has torn apart or making new kidding pens. Once after helping me with a carpentry job, he finished the barn cleaning I had started when I was exhausted from three weeks of goat birthing with more births yet to go. John's given me good hay too dusty to feed his horses; he once gave me a winter's supply of cedar kindling when my fire-making skills were hardly reliable. Another year he gave me a load of small logs for stove wood. And I know I can call him in the middle of the night to help with a difficult birth. So far I haven't had to, but I know the offer's still good.

Because I've tied myself down with goats and dogs and cats and therefore seldom leave the island, people call me to look after their animals when they're away. Glen and Kim have repaid me with many favors. Glen helps me hold the billy when I must take his temperature or give him shots. He does it cheerfully, and has about him a way that is accepting and affirming. Kim also helps and brings me surprises from the mainland—artichokes or tropical fruit. I don't want to refuse to help these friends. It's part of the beauty of life on the island.

Snow keeps falling. It's now been forty hours and there's no sign of letting up. Last night I watched the local weather report, though "local" is ninety miles away in Green Bay. I seldom watch it. Even the closer Sturgeon Bay radio station is not very local to this little island and the weather news seldom applies. Last night it said the snow had stopped except for occasional snow squalls over the lake. The snow here continues, steady and thick.

John comes by to give me last-minute instructions and numbers to call if Gunner, one of the older horses, has a relapse of last year's illness. John says the heater for the stock tank doesn't always work but that the horses can eat snow if their water freezes. I write letters and clean the house, then lug two toboggan loads of firewood from behind the goat barn to the wood box. When the snowfall is moderate, I shovel a path for the wheelbarrow, which holds more wood than the year, changing the rules? By seven there's faint light. I climb into insulated coveralls and boots and go out to check the state of the driveway. The van warms, motor rough, while I feed the goats. It's twenty degrees in the barn, so probably it's about eight outside. Snow has blown in the doors to the goat pens but the animals seem unconcerned. I brush the snow off some of them and wake up others to make sure they're alert and well. The barn skylights are covered over with snow so it's darker in the barn than usual; the goats appear sleepy, like children not ready to be awake.

I attempt the driveway. The snow is so light and powdery that the van succeeds despite the accumulation. Later I try snowshoes again when I go out with Lily, but they sink deeper than yesterday. I go back for cross-country skis and proceed in Lily's path. The snow is up to her back, a good seventeen inches, but she leaps out of it, then plunges in again. We cross two fields and part of another this way, then turn back. By this time the snow has stopped and patches of blue show in the western sky. The sixty-hour snowfall has ended.

Navidad

THE NORTHWEST wind is an honest adversary. Sharp, relentless, it tells you up front what to expect: You know the ferry to the mainland may not run when it blows; wintertime, you park the car behind the house, out of its path, and expect the woodstove to take twice as long to heat the room. Unlike the north wind, the south wind is devious; it's not what it seems. Its thick edge suggests mildness, hints of spring even, and then it sweeps in with a wet, heavy breath that's somehow unclean. Goat kids sicken and die when the south wind blows.

I find her leaning against the creep pen, shivering, head tilted back, eyes half closed. Yesterday she scampered across the barn floor, ricocheted off the grain trough, danced into mounds of alfalfa, and tapped her tiny feet against a metal feeding pan as if beating a rhythm for her own amusement. To watch her is to hear miniature silver bells, as finely made as her translucent infant hooves, tinkling in the sunlight. Today she is mute; all I hear is my own silent cry.

Late yesterday the south wind began to blow. It surged across the swept tundra of gray-brown stubble, February's pasture. It coated every fence post, every frozen stick of knap-weed and quack grass with its viscous rawness. It invaded the barn through a low portal, frozen open until spring loosens the door's cement-block doorstop. It clung to the warm-blooded animals, spreading over them and penetrating them despite their thick, curling fleece. The little Angora goat, born too early, is no match for such a force.

She's still very young and her hair covers thinly. In damp weather, sparse curls the color of antique satin lie plastered against pink skin; on dry days, her coat fuzzes to cotton candy. At night she sleeps nestled low against her mother, the doe shielding her from drafts. Ideally, she would have landed at six and a half weeks of age well on the other side of the vernal equinox, when a sun that truly warms balances the chill rains and occasional snow of April and May. The internal thermo-stat of a baby goat normally struggles with the pneumonia-inducing vagaries of spring temperatures. However, the fluctuations of weather challenging this small kid range from a breath-stopping Arctic cold to the gelid rawness that presages a temporary thaw. She won't nibble stray blades of green grass for another month or two; pastures on Washington Island aren't lush until even later, nearly June.

One glance at her and I feel something in the area of my stomach shift into near-panic gear, adrenaline fueling it. Automatically I turn to set up an enclosure for her in a room we've just insulated for use as a maternity area. With painted walls and an electric wall heater, it's luxurious compared with the rough, drafty area where the goats kidded last year. There, water froze regularly in plastic drinking pails set in high-sided maternity pens. At four o'clock one morning I restacked fifty or more bales of hay against the west wall to try to stop a strong wind from gusting through openings between weath-ered boards. At dawn, I tacked up plywood outside. Dangling from rafters, heat lamps warmed newborns on nights too bitter for merely the mothers' body heat to suffice.

I spread hay thickly on the cement floor within the confines of the four-foot-square hardboard pen. I scoop up the baby, who does not resist. I want to hold her to me, give her warmth and comfort, but she's all legs, gangly and awkward in my arms, and shows no desire for my mothering. I position her on the hay, then go back for the doe, who hovers by the door. Normally she stubbornly resists any attempt to lead her, but this time she walks along willingly as if she knows she is rejoining her baby.

I dab petroleum jelly on a thermometer and take the baby's temperature. The silver band streaks past the last calibration, one hundred and six degrees. An adult goat's normal temperature is one hundred and two; a kid's can be as high as one hundred and three. If a fever of one hundred and six persists, neurological damage threatens.

Named Navidad because she was born December 23, she scoured, or had diarrhea, and a slight fever at one month. I felt lucky. The temperature indicated a bacterial infection rather than coccidia, a parasite that lodges, sometimes fatally, in the intestines of kids, rendering them incapable of absorbing nutrients. Called the "silent killer," coccidiosis stunts the growth of goats that don't die first. I treated her with penicillin injections twice a day for five days. She never seemed sick and continued to dart about the kidding pen where I confined her with her mother. She sampled hay and nibbled at grain and took long drafts of milk from the doe.

Two weeks later she does not scamper away when I reach down to pick her up, nor does she try to wriggle from my grasp when I take her temperature. She leans weakly in the corner, shaking. Her head jerks oddly from side to side like a person suffering with palsy. There's no diarrhea, no runny nose, no swollen joints that I can detect.

The wall heater fills the room with its distinctive odor of dusty electrical elements warming up, but the air remains chill. I rig a heating lamp for her pen by improvising hooks out of two nails pounded into the ceiling at a slant, crossing midway to hold the lamp's cord. I place the little goat directly under the heat, about four and a half feet away, but she falls back into a depression in the corner. After a last searching look for clues to her illness, I call the vet.

With the winter's ferry schedule of only one boat a day, there are no impromptu trips to the peninsula until later in the spring. Veterinary emergencies are waited out or resolved as best they can be by telephone.

The doctors do not presume to diagnose long-distance. With no examinations possible, no blood analysis, no fecal sample for scrutiny under the microscope, they listen to their clients' descriptions of problems and suggest a therapy. It must be a little like old-time country medicine, when a doctor prescribed on the basis of his observations and his knowledge of what served before. But here, the veterinarians must depend on secondhand, unschooled reporting, related by telephone. The most precise information they work with is the reading on the thermometer.

The receptionist puts me through to one of the doctors out on his rounds who has just called in. It's the same vet who guided me through Navidad's last illness. He listens silently, then makes brief suggestions, combining economy of words with unmistakable compassion in his voice. Could I get some Banamine into her right away to reduce the fever? I must start her on penicillin, then try to find some sulfa because she was on penicillin earlier. She must have liquids. The office will send the medicine I need, but the mail takes two days, sometimes longer, so the vet suggests people who might have some I can borrow in the meantime.

I pick up the telephone and start dialing the people on the list whom I know, then I'll try the ones I've never met. I don't hesitate. Islanders pull together willingly, gladly, not only in emergencies but at other times, too. To choose to live cut off from most of the world by six miles of icy, surging water is to forge an instant kinship with the six hundred or more people who live here year-round. Perhaps it's because we see a part of ourselves in every other inhabitant, no matter how unlike us they may be in other ways. What we recognize is that part of ourselves that prefers solitude to convenience, a simple life to luxury, that thumbs its nose at the perquisites of twentieth-century American consumerism. It's a recognition that kindles unquestioning generosity when it comes to surmounting difficulties posed by island living.

First I call John, who has Banamine in case his horses develop swollen joints. His Banamine is out-of-date. I telephone Evy. She has it but is low on the sulfa I need, which she's giving to one of her horses. Perhaps Lois who was treating a horse that died recently? I reach Lois, but she has already discarded all of Barnaby's medication. Perhaps Karen? But no, she doesn't have any.

Outside the wind has shifted, now lashing the house and barns with frigid blows from the north. The air is noticeably colder but also less damp. I take the wind personally and rail at it mentally as I trudge through the snow with a syringe containing a mere half cc, or one-tenth of a teaspoon, of penicillin.

I draw a sturdy, paint-splotched chair to the kidding pen, then lift the baby onto my lap, turning her small limp body so that I can grasp the muscle on her hind leg with the thumb and forefinger of my left hand while injecting her with my right. There's so little flesh on her leg that a worry flickers momentarily across my mind that I'll injure her. What if I strike a bone? I dismiss my concern and note that at least she reacts to the needle with ever so slight a wince. A good sign, I hope. All the while, the doe stands near the side of the pen, reaching over occasionally to pull at my hair and jacket with her mouth.

In my van, driving to Evy's for the Banamine, I will the kid to live. It's a will that's stubborn and unyielding. It energizes me, circulating intention to my cells and lending me a power that may well be illusory. However invincible it feels, I know my strength of will affects the little goat's fate only slightly. All my determination is only a weak cry in a windstorm.

I navigate the long icy driveway at Evy's and turn the van so that I can exit easily. I don't want to waste a minute getting Banamine into the baby and must take care to avoid having to extricate myself from a snowbank. At the door Evy hands me not only the bottle of the clear liquid but also a small plastic container holding more than enough for three days of the pink sulfa medication.

Driving back I notice the sky has darkened; a shine from car headlights skips over the iced road. Wind claws at the van, and I remove my gloves to grip the steering wheel more tightly in order to steady the vehicle on the slick pavement. I transform my concern for the goat into frustration at the elements that seem bent on impeding me. The trip takes doubly long, or so it seems.

Once home, I rummage in the cupboard for clean syringes and an appropriate needle, then crunch through crusted snow to the goat barn. My hope for the kid's recovery, momentarily buoyed to near euphoria by the medicine in hand, fades quickly at the sight of Navidad. Her hind legs have stiffened and her front legs collapse at the knees with each step she attempts. I search inwardly. Where is my resolve? I'm not ready to accept losing her. Yet common sense predicts I'll see the death of many kids as long as I raise goats. I inject her with Banamine and squirt one and a half cc's of the thick cherry-flavored sulfa liquid into her mouth, then turn my attention to the question of whether she's getting enough fluids.

The doe's udder is full, a sign that the little goat has not been nursing. The mother is not an easy animal. We bought her four months ago from a woman who admits she prefers her passive sheep over her inquisitive goats and gives the goats less attention. Milking the skittish mother will not be easy. I climb into the pen and work the doe to the side opposite the kid so the baby won't be stepped on. I shove a large pail under the udder, hold the old female by the horns with one hand, and press high on her teat with my thumb and two fingers to encourage a stream of milk. After several moments, the doe lurches away from me. I try again and manage to coax out a few squirts, but never the flow of milk I'd like. I persist until I have roughly an ounce, which I suck up in minuscule amounts into a three-cc syringe to force between the baby's clenched gums. As I do so, her eyes close and she gurgles in the back of her throat. Finally, she refuses any more by letting her head drop limply to the side.

I've nearly forgotten the other goats. Some are clustered in small clumps of two or three outside the barn; others mill about inside, waiting for their evening grain. Their fleece is long, about six inches now that it's just before shearing. The mohair's oil, which helps protect the goats in wet, cold weather, has picked up five and a half months' worth of dust and grime. It's as gray and dingy as the trodden, manure-soiled snow on the south side of the barn where they like to lounge. The does are in their last month of pregnancy, and they're ravenous for their mixture of oats, corn and soybeans laced with molasses and a vitamin-mineral additive. I feed them and note that everyone is eating with healthy greed. Later, the hugely expectant does will settle themselves in the hay, looking like Spanish galleons plying the waves.

Attending the apparently well adults and yearlings clears my mind. Those twenty minutes have steadied me and given me a strengthening respite. Yet part of me knows I can't allow myself to regard the little sick goat with perspective. At least not yet. Until she's either better or dead, I'll continue in high gear, in a mode of intensity.

The next steps align themselves in order of importance. I'll try Pedialyte. The island's one grocery store carries it, and there's just time to get some before the market closes. Later, I'll try to milk the mother, and if I fail, I have powdered milk, kid milk replacer for orphaned goats.

At eight p.m. I report in to the doctor. Navidad has swallowed, however unwillingly, at least two ounces of electrolyte solution. I wiped another ounce or so off her mouth and my jacket. She's also taken a couple of ounces of milk, and I'm prepared to force more into her every few hours. She still shivers but her legs have stopped buckling, and most significant, her temperature has dropped to one hundred and four.

I won't let down my guard yet. I won't feel the cold when I make my way across the snow to the barn to check on her at ten p.m., at midnight, and at two a.m. I'll fret as I feed her, worry that she's not taking more. But now I can feel Navidad's private storm abating, and outside, as if in concert, the wind is lessening, too.

More Kids

DURING MARCH if the snow holds I can still count on cross-country skiing through the woods with my dogs. But it's taken a few years to achieve this, which all hinges on the goats giving birth conveniently in April. Aside from Noel and Navidad, the second crop of kids began arriving March 17, a cold night when the temperature plunged to zero. Glancing around our splendid kidding room, walls lined with particleboard that I'd painted a soft peach only a month earlier, I felt secure in the face of the weather. And to combat severe conditions—minus thirty or more—there was an electric wall heater in the room. John had covered the seams of the sheets of particleboard on the ceiling and walls with strips of wood, which were just adequate for supporting the nails I pounded in to serve as hooks to suspend heating lamps. I borrowed from Jeannine a one-way baby monitor that allowed me to hear sounds in the barn. For the duration of the kidding, I lived with its constant buzzing, night and day, so that I could be alerted by the sounds of goats in labor.

My hands shook with the first birth or two, but soon I relaxed into it, felt competent and could enjoy the overwhelming sense of the miraculous that accompanies each birth. By the fifth day there were eight infants, some small enough to need further confinement with their mothers. Two days later the number had more than doubled to nineteen. The kidding room and an adjacent room, slated to serve eventually as a veterinary room, were full. I moved mother and baby pairs out when I could and cleaned the pens, trundling dirty hay and manure by wheelbarrow over the snow-covered, ice-encrusted lawn to a field where I dumped it. On March 22 I began setting up kidding pens in the hay storage part of the barn, using pallets tied together with baling twine and draped with burlap feed sacks to shield the goats from drafts. I strung heating lamps from the rafters and bolstered the pallets with hay bales. My goal was always to keep two spare kidding pens ready at all hours, so I often found myself moving goats and cleaning away soiled hay in the middle of the night.

At noon on the twenty-second Gilda bellowed her labor. I had every expectation that this year's birth would be uneventful, unlike the previous year when her stillborn baby was our second kid born. A half hour passed without sign of a kid. I felt inside Gilda for the baby and could barely touch it as it began its journey in a birth canal that was too long and too narrow. I called Donna to help and together we worked for an hour to deliver a female kid. Exhausted from the ordeal, Gilda had no energy to lick and nuzzle the baby, so I dried her with towels and unwittingly allowed her to bond with me to the point that she refused any milk from her mother. This sent me to the refrigerator in the house for spare colostrum, then back to the barn to feed the small, weak creature by dribbling warm milk into the side of her mouth, one cc, or one-fifth of a teaspoon, at a time.

During this time another kid was born, easily enough, and another at four-thirty. Close to midnight I heard sounds of a goat in labor. I rushed to the barn and found Elsa, one of our original Texas does, with two kids in the hay in the does' area. I grabbed a clean towel and scooped them up, one wet baby at a time, and carried them to one of the kidding pens in the frigid hay-storage area. Then I guided Elsa to the pen so that she could resume licking the babies dry. Just as I was clapping a jar of iodine over the umbilical cord and navel of the kids, I heard the regularly spaced bellows of another goat in labor. I finished more or less with Elsa, trusting that she could get her babies started while I checked the newest mother, Amneris. Her bellowing grew louder, her straining intensified. With a hollow feeling, I realized I could see the baby's head but no hooves. I switched a goat and kid from the maternity area into a new pen to make room for Amneris, then I dragged her into the kidding room, where the light was better. As one of the newer additions to our herd, she was still wary of me and clearly disliked my intrusion. I penned her, ran back to the house to scrub my hands, greased them and began feeling around the baby's warm, slimy head and shoulders for a leg. By pushing the head back into the doe, I could work my hand in to grasp a foreleg that was pinned next to the little body. I brought it forward but could not manage to find the second. One would have to do. I tugged gently with each contraction, working the kid's shoulders through the opening. Astonishingly, a wet, yellowish body slid onto my lap. I put it under Amneris' nose and left the infant and mother alone for a few moments to establish their bond.

Stepping from the brightly lighted maternity room into the darker, colder hay–storage area, I found both of Elsa's babies lying flat on their sides, not moving. I lifted them to Elsa's teats, one at a time. The little female showed interest, but the male hardly responded. Worse, the inside of his mouth was cold. Gilda's baby still needed attention, and Amneris and her baby needed routine care plus a penicillin shot for the mother. It was one in the morning and I needed help.

Toni had said to call any time of the day or night if I need-ed her, and she came immediately, within fifteen minutes, from the home she and Gene have on the other side of the island. Having no livestock experience whatsoever, Toni set to work. We took turns holding each of the twins, encouraging them to drink drops of colostrum. Toni rubbed them with a bath towel and continued to dribble milk into them while I attended to Gilda's baby and to Amneris. When I was free I gave Elsa's male baby shots of dextrose. It was cold in the barn, and raw; bundled in heavy jackets and scarves, Toni and I could see our breath in the air. After a couple of hours there was no more that Toni could do. Gilda's daughter had taken adequate milk from me for the moment and Gilda seemed to be resting. Amneris' kid had suckled and was nestled next to his mother; Elsa's daughter seemed out of danger. Only the little boy was failing to take hold of life. At somewhat past three I walked Toni to her car, then carried the little male into the house. I lined a cardboard box with towels and rigged up a heating lamp in the dining room, safely away from the dogs.

I had not used a stomach feeding tube before, although I had one on hand, ordered the year before for emergencies. I dreaded using it; I was well aware that the first time Donna used one she inserted the tube down the trachea instead of the esophagus and pumped milk into a newborn lamb's lungs, drowning it instantly. I removed the apparatus from the cup-board and began reading accompanying instructions. I glanced at the clock: three-thirty. Fleetingly a sense of the bizarre seized me; nothing in my life up to coming to the island could have predicted this early morning scene. I held the leaflet, reading and rereading it. I hated to attempt using the feeding tube but the kid would surely die if I didn't get milk into him in some way. Any milk I attempted to give him with a syringe had just bubbled around his mouth and dribbled down his chin and neck. After collecting some colostrum from Elsa, I locked the inquisitive dogs away and brought the kid into the kitchen, where the light is better. I drew warm liquid into the two-ounce syringe that attached to the feeding tube, then put it aside, hoping it wouldn't cool too much. The instructions said to insert the tube eight to ten inches into the animal, so I took a ruler and pen—this procedure seemed too critical to chance a guess—and marked the eight-inch point. Then I wet the tube and put it into his mouth, slowly forcing it down his throat. It went only so far. I removed it and tried again. The kid was too weak to resist, and the second time with my hand on his neck I could feel the bulge of the tube as it went down his throat. Once in, the hard part was keeping it down with one hand while I fumbled to attach the syringe with my other hand. I managed, although I lost some of the precious liquid. Then slowly I pushed the plunger, forcing milk into the little goat's stomach. I was prepared for instant death, but it was instant life that I saw, for immediately he became alert.

I didn't sleep at all that night nor the next day. Daylight hours blended crazily with the night as I drew energy from the goats and the excitement of birthing. I tried feeding the failing twin with a syringe but had to resort again to the stomach tube. His mouth went cold between feedings and he lay limply on his side. But after a little milk he would raise his head and look around. He was a large kid but looked oddly narrow through the chest, as if it lagged in development. During the next two days eight more kids were born but, conveniently, they came during daylight hours. I kept the heat lamp on Elsa's baby and fed him with the tube, but he did not thrive. Around noon on March 25 he died.

By that time Elsa's udder was becoming engorged with the milk she was producing for two. I felt hard places and knew it was an early sign of mastitis, an inflammation of the mammary glad caused by bacteria. Mastitis can mean pain and fever and ultimately one side of the udder turning black and gangrenous and sloughing off. I started penicillin shots, which Elsa accepted with dignity as long as I did not show undue interest in her baby. After having one twin swept from her, it appeared her main concern was to shield her little female from me. Eventually her milk flow adjusted, the threat of mastitis subsided and she continued mothering her daughter with seeming affection.

Thirty-two kids were born in three weeks' time that year. Gilda's little female thrived on bottles but cuddled next to her mother at night and allowed Gilda to clean her and watch over her. Another bottle baby was the result of an inexperienced mother who rejected her offspring at first. The baby was so tiny initially, and would not suckle, so I had to feed her four times with a stomach tube and then at two-hour intervals round the clock during the first few days. For lack of space I kept her with her mother, however, and gradually the doe accepted her, although like Gilda, she no longer had milk for the kid. At least two more first-time mothers lagged in the development of maternal instincts. One of them bolted from the barn, where she had just given birth, and I had to chase her to bring her back to her baby; the other butted the helpless newborn. But within a few hours, both does accepted their babies. As time passed, it was fun to watch the antics of the kids newly released from confinement. And as the weather warmed, and the goats ambled to the far end of the pasture, I'd note that one or two does—often Elsa—would stay behind to watch over the kids ,who generally stayed close to the barn.

It was both a heady, invigorating experience and an exhausting one, this goat birthing interlude. I sometimes felt like a waitress during the high tourist season in Door County, as I dispensed clean water, hay, and individual pails of grain to the goats in nineteen kidding pens, cleaned up after them, and changed their bedding. The work, though, was reward enough in itself, for it awed me with a glimpse of something ancient and miraculous and yet as commonplace and elemental as the dawn of each new day. I never lost a sense of wonder and enchantment at the babies taking on life, metamorphosing before my eyes, it seemed, from inert fetus to a quickened being as they struggled to stand and groped for their source of milk, while the mothers cooed a goat lullaby of soft little grunts. I would lose in those moments any sense of otherness, of any difference between animals and humans.

Bittersweet Summer

AT MY BROTHER'S farm the lawn sweeps down from the shingled, two-story house stationed majestically on its rise of land, runs across like surf on a beach to the barns, encircles them, then spreads down to the road. Over the years Peter has enlarged the lawn, creating new turf by simply mowing what had been rough tracts of adjacent fields.

A careful husbandman who prides himself on neatness, he has mown broad borders on either side of the road and has planted two long rows of pine trees on these strips of grass. They serve to stamp the land as belonging to a person who likes his property well defined and orderly. It falls to me most of the time to cut the grass on the two and a half to three acres of lawn, and I do this with one or the other of the two lawn tractors my brother keeps here. Because the rocky ground is uneven and the tractors jolt and bounce, they are continually being jarred into disrepair. But one or the other will be standing by, ready to fill in. There are certain areas, too, that need the attention of a gas-powered push mower and when that chore is done, I use a trimmer to cut the errant blades around the house, pump house, fences, and, when I'm feeling ambitious, the barns.

I've grown to enjoy these sessions on the lawn; there is a feeling of bonding with the property, of knowing every square inch and its idiosyncrasies. In mid-May, when I start cutting, it's likely to be cold and it's easy to feel chilled while astride the lawn tractor. But by mid-June, summer is coming into its own and it's sweet to spend a long evening working out of doors until nine and then, as I did all one year, take a quiet drive on silver ribbons of deserted road to the west shore to watch the sun setting over the water. It's no wonder that the island was settled by Scandinavians and reputedly at one time was home to the largest group of Icelanders in the United States. The long nights suddenly shorten in late spring and provide an exhilarating release that must be similar to the abandon and freedom of the summer solstice in Norway and Sweden. Then I cannot be out of doors enough, from the first light of dawn around four to the last glimpse of light in the sky at ten. Suddenly, too, it's as if nature is working double-time. Spring flowers that bloom in sequence elsewhere burst forth simultaneously here. Tentative, tender spring green gives way almost precipitously to full-leafed summer splendor as if there's not a moment to squander in this short season.

Early summer, with its honeyed light and long twilight's luxuriant darkening shadows on emerald grass, always seems a reward for getting through the hard months, the months of Arctic cold and ice and bitter wind. The goats spend most of their time outside then, which makes barn cleaning light work indeed. They eat mainly pasture as long as it is green from sufficient rain, and so work shrinks to a minimum, mostly hoof trimming, delousing, and worming. In the summer I feel most the lack of a companion. As long as there is light I keep myself busy with gardening and goat work and farm and lawn maintenance, all of which I enjoy. But I know, too, that if I had someone to share the moment, at day's end I would choose to sit out on the lawn, perhaps a glass of wine in hand, to enjoy the peacefulness and beauty of the soft evening air, goat kids frolicking, whippoorwills down by the swamp beginning their three-note call.

My summer idyll on the farm ended the year after the second group of kids had begun to venture to the far end of the pasture with their mothers. Peter wanted me to move off the farm for the summer so that Jeannine and my nieces could use the house, and an artist friend could use the guest apartment and the studio. "It's our place and it's only fair that we get to use it for six weeks or so in summer," he said. And he was right. But I felt jolted by the necessity of uprooting myself and finding other quarters and was dismayed but not surprised at the prices for summer rentals. After I had searched in vain for something reasonably priced, Toni and Gene kindly offered a large spare room over their garage. Peter agreed to pay them a small rent, and I moved in with my dogs and cats.

That summer my niece Anne was six and Jessie, born the previous August, not yet a year, which gave Jeannine enough to do without having to care for goats. So I came over twice a day to check on the herd and make sure they had feed and clean water and to do routine chores. And I helped with the haying, the majority of which was not done at Peter's farm. It was wrenching to leave the goats and the farm because I loved being there, where I felt I had a purpose and an identity. The difficulty was compounded, as well, by feeling that Peter and Jeannine wanted their life to themselves on the island, and this made me hesitant to intrude. I spent three summers with Toni and Gene at their house on a point of land that juts into a small bay. Later, when their spare room was occupied, I stayed in the woods overlooking the water on the island's north side, in a summer cabin owned by Jim, a friend of a friend who does not use it much. These places were beautiful and gave me a glimpse of other parts of the island, but still the summers meant a feeling of loss.

Earlier in the year, the Polywire fence had begun breaking down. By early summer, before I left the farm, the goats broke through the fence regularly in order to sample the grass elsewhere. Once I saw a car stop to let the herd cross the road; I ran out to apologize and thank the people, who turned out to be good-natured and much amused by the incident. Once I'd contained the goats after a breakout, I'd turn off the fence charger, search the perimeter for faulty wires, which I'd fix, then plug in the charger again and test the fence with a voltmeter. By that time, the goats would have escaped again through another section of fence. None of this seemed to faze my brother, although I complained long about the problem. Finally I reported that the goats were enjoying their nibbles of his blue spruce. Within days he had engaged a fence company representative whose brochures I had picked up at Fiberfest the year before. The fence man stayed three days; he and Peter and Peter's friend Bill enclosed a new pasture of thirty acres with a good, strong high-tensile electrified fence. It was arduous work because of the rocky soil, but the result was a virtually trouble-free fence that endures beautifully, even standing the test of a lust-driven billy goat.

Autumn

IN SEPTEMBER the sun shifts perceptibly, creeping from its summer vantage point in the north; then one dawn near the end of the month I'll look out an upstairs window and see it rising to the right of due east and I imagine it pulling autumn along in its wake. In August an aberrant frost slicked the barn roof; it was a brief foretelling of the long, dark cold I want to banish from my consciousness. But now in early September, it's summer again, lush days of ripening tomatoes and thick slashes of orange and yellow and rose in the flower garden where mature annuals are finally coming into their full-blossomed best. Thunderstorms keep the lawn emerald and growing faster than I would like. But also the pasture greens again after a dry August and this pleases me, for it gives the goats another month before I have to dip into the supply of hay put up for winter fodder. Full-coated like the late-season flowers, they stand out, a string of bulky, creamy tufts bunched here and there, then trailing out against a verdant carpet.

Against this serene backdrop I try to put my life back together, try to establish order, get the chores behind me. But as always, I fail in September, as if the shifting seasons won't let go of me in their effort to remind me that the world is in flux and so must I be. Summer will not last, I'm reminded, but also I'm told that in the face of nature's great changes, the shifting of the order, the yearly vernal-autumnal cataclysm that propels us from ease and warmth into struggle and ice, I am not allowed to turn my mind away to other pursuits. I must somehow participate in nature's upheaval. My attention span shortens with the curtailing of daylight hours. I turn from one task to another, restless and never quite satisfied. I work in the garden, convincing myself that it's still summer while snipping dry stalks of July's delphiniums and the daisies that remain fresh most of August in this chilly climate. It's not nearly time to put the garden to bed, yet my cultivating and weeding and transplanting now seem like idle play. I give up and turn to mowing the grass, careful each time not to cut the lawn too short and thus vulnerable to a sudden hard frost. I'm most content in this tumultuous season when I have a defined task before me to steady my thoughts and circumscribe them lest I plunge into dreams of new challenges unsuited to life with three dogs, four cats, and ninety goats.

Every year in September I think now is the time I shall get back to my desk, to my writing. August is over; Labor Day ends official summer on the island although most vacationing families with children have already left to start the school year elsewhere. Suddenly the island is quiet, the population pared from a summer peak of some two thousand down to six hundred fifty. The road shimmers in the twilight, solitary and untouched in the evening hours except by the hooves of the white-tailed deer that begin to show themselves again. The drive-in restaurant closes except for weekends. Sunset Resort still takes guests but does not serve breakfast weekdays. The parking area at the market empties and remains uncrowded most of the week; in the deli section at the store the case of pasta and bean salads, cold well-done roast beef and ham, baby Swiss and Muenster and cheddar is cleared out and a Closed sign tacked up. Islanders give a mental sigh and gear down; the short time to make money is over and although money is a necessity, it's never been the priority here that it is other places. Islanders prefer to set their own pace, which includes ample time to visit and gossip, to help each other, to fish and to drive the roads after dark to scout the fields for deer in anticipation of hunting season.

During the short tourist season and into autumn people here work hard, whether waiting table or cooking at the local restaurants or doing carpentry and repairs for the vacation houses of summer people. Our friend John is often up at four to work on one job before going to another at seven; later he puts in a few more hours at still a third after four-thirty in the afternoon. It may be a new deck for one person, insulation for another or the remodeling of a house for a third. Only then does he go home to work in his vegetable garden, often at the same time smoking salmon for sport fishermen. Mack crowds his schedule similarly. He and his brother own the local liquor store, which sells his mother's pies, coffee cakes, and cookies in the summer as well as moccasins and television sets. But Mack also has a transport business, hauling loads on the mainland, and an electrical repair business. In summer Mack goes from one job to another fixing refrigerators, dishwashers, and VCRs or installing satellite dishes or washing machines. When I call Mack for a repair, I specify whether it's critical, because I know if I leave word that the job is not an emergency, he'll be there all the same the next day at the latest. Mack doesn't stop in the summer, although the perception among summer people is that the island is a very laid-back place. This perception originates, I suspect, in the unfailing way John and Mack and others conduct themselves with quiet courtesy and cheer. They never seem rushed or harried yet they do as much or more than their counterparts in the city, and this gentlemanly demeanor, born of inner peace and self-confidence, I like to think, must be misinterpreted by some. Another example is Dave, who has the only gas station and automotive repair business on the island, servicing nearly every vehicle in one way or another as well as lawn tractors, power mowers, snowmobiles, and some boats. For all the volume of work, Dave and his mechanics respond immediately to emergencies and still have time to deliver a repaired car and chat for a moment. One recent evening well after six, with his two youngest sons in tow, Dave drove up pulling a wagon carrying the repaired lawn tractor and the gasoline-powered push mower. It was getting late, but he had time for a discussion of the state of development on the island, then the four of us strolled to the barn so that his small boys could feed grain to the goat kids.

In late September a visitor stopped in the goat pasture. It's been a lazy autumn without early blasts of north wind that rip the trees bare of leaves. I pause often at this time of year to absorb the changing colors of the sugar maples and birch that fringe the goat pasture, hoping to create indelible pictures in my mind. During one of these long gazes I noticed a dark smudge against the newly green grass, verdant after a fall drenching. I walked into the pasture and without having to approach too closely saw that a Canada goose was settled on the ground about a third of the way down the nearly quarter-mile fence line. Its black neck rose in a graceful line but its bill pointed downward. Was it sick or hurt? I wondered. I walked toward it, and it rose easily but without hurry and flapped its wings a couple of times. It did not take flight but nevertheless moved several yards farther into the pasture, where it settled again. I checked it occasionally during the afternoon and later brought the binoculars up to the bedroom so that I could observe it from the window.

The next day she had moved back to the fence line where I first noticed her. There is no difference in appearance between the sexes in Canada geese, but I felt our visitor was female, perhaps because, like Penelope awaiting Odysseus, she waited so patiently. I noted that she flapped her wings to pull herself into a standing position only when I moved toward her. Was she weak from illness? Would I soon dig another grave alongside the new one of the kid that died recently? Was she hungry? I dipped into our feed mixture of corn, oats, and soybean meal, molasses and vitamins. Geese like corn. Farmers complain often of the copious and rather liquid droppings of geese in their cornfields. I penned the goats in the barn so they would not see me distribute a treat in the pasture, then sprinkled the grain along both sides of the fence far enough into the enclosed field so that the goats, once released, would not find it immediately. She moved away but I hoped she would dine that evening before the kids and does inevitably discovered the feed.

The next morning at first light I grabbed the binoculars and trained them on the pasture. She wasn't there. I felt heavy with disappointment, and I was puzzled. Surely she had not left during the night. Later, as the mist that often hangs low over the ground in the fall burned away, I saw the dark shape settled on the boys' side of the fence. It was unlikely that she had slipped through the electrified fence wires, so she must have flown over. Reassured somewhat that she could indeed fly, I still worried about her. Mid-morning I heard the honk-ing of geese and looked up to find a flock flying in formation overhead. I watched our visitor but she showed no interest, did not even raise her head. Later I washed off a large blue plastic tub, the kind that comes with a lid and is meant for storing food or clothing. I bought it as an emergency water vessel when below-zero temperatures froze the water solid in the goats' automatic waterers. I hauled it out to the fence and poured water into it. The goose had moved off by this time but not too far, and later I saw her settled near the tub.

Another day passed. I would see her next to either of the two interior fences, but most often by the water tub. I sprin-kled grain just outside the pasture and also by the tub, which was a mistake because the wethers and billies soon noticed the feed and displaced the goose while they gobbled it. I began to think of her as a benevolent if puzzling fixture in the pasture and wondered if she could overwinter here. A few Canada geese make a home in winter at the Chicago Botanic Garden, so why not here? On the fourth morning of her visit I checked her early as usual before feeding the goats, and this time there was not a single goose by the tub but rather two. An hour later both were up pecking at the ground. By mid-morning they stood outside the fence exactly where I had strewn feed the day before. I was much relieved that she now had a compan-ion, surely a gander, and I was relieved, too, when they van-ished the next day, most likely off to wintering grounds in Louisiana or Texas or even Mexico. I liked it that she appar-ently sensed a haven here. And I am awed once again by all we cannot know about nature. Was she waiting for her mate? Was our farm a prearranged meeting site? Or had she become separated from her life partner and simply waited until his instinct homed him to her? Or could it be that she lost a mate to accident or illness and had decided not to make the long trip south until another solitary bird spied and courted her? I hope they've joined a flock and are headed in V-formation to their winter home. I will never know the answers, but the questions will nag until I make time to search the library.

Shifting Wind

SOMETIMES I felt as if a lens moved over my world to distort it slightly. Nothing had really changed, I told myself, but my perceptions kept noting differences. A change was in the air. Peter's enthusiastic weekly telephone calls had long since dwindled to only occasional communications, and then not much was ever discussed. Where, I wondered, was the past spirit of adventure? What was the next step with the goats, and where were we going with them? I ventured to bring up the subject on various occasions but to no avail. "It's not something I can discuss at this time," Peter would say. It was futile to press the subject. Peter could and would talk long and eloquently when he was ready, but no gain was ever made by insistence when he was not. Once when I lived for a winter in a small hotel in Paris, fleas bit me every night. I wrote asking Peter's advice because he was bitten by fleas when he was in the Peace Corps. He never answered my query. Later I asked him why. "There was no point," he said. "There was nothing you could have done about them anyway." There was no longer talk of the four hundred wethers and enough does to keep up the number. That had been a joke—partial joke—of course, but it had a jaunty, optimistic ring, full of youthful bravado and dreams.

Just a year after we had bought more goats Peter announced he wanted me to start selling some of our animals. It was true, they were flourishing, the herd growing perhaps more quickly than he had anticipated. Little Bill, Lucy's son, whom I'd kept as a billy, was the first to go. I reasoned that we did not need a billy who was half-brother to many of our females. A show farm and petting zoo in Sturgeon Bay needed an Angora buck so I took the ferry across and drove there to inspect the facility. It was beautiful and very clean and the animals all appeared sleek and well fed, with the hooves of the goats nicely trimmed. I made the ferry trip once again to deliver Little Bill, and cried as I drove away. In October I sold four does to a woman in the western part of the state. I picked out females who grew heavy, curly fleece but for whom I had never developed much affection. I loaded them into the back of the van and drove them ninety miles to Green Bay, where I met their ride to the far side of the state.

Peter did not want kids the next year, he said, but he had not yet added interior fencing to the splendid new perimeter fence that enclosed the thirty-acre pasture. So there was no means of separating the does from the billy goats.

"But I thought they didn't breed until November," Peter said when I told him we'd surely have kids. I explained that November was merely the time I introduced the billy to the girls; left to their own devices they could breed as early as August. And because offspring were assured despite Peter's wishes, I began breeding some of my favorites selectively. I'd notice when a doe was ready and put her in a separate area with José, who was the better of our two bucks. In time, Peter and my nephew Bill put up interior fencing—in the bitter cold and wind and clearly too late to prevent unwanted breeding.

It was apparent that birthing would start before the weather warmed enough to risk shearing the goats. The hugely pregnant females would need to be crotched—the hair taken off their bellies and the inside of their legs so that newborn kids could find a teat easily. I enlisted a friend to help me. She held the does to steady them for me—and steadied me in doing so. I had not taken up a shearing machine for more than a year and felt nervous about it at first. Very soon afterwards, on February 9, when the temperature was minus ten, the first wave of what would be twenty-seven kids was born. The kids came quickly and with the bitter weather I dared not leave the farm lest a goat freeze before I found it. When I finally had to give in and make a trip to the grocery store, I enlisted Kurt, whom I hardly knew, to stop by and listen for newborns. That is the beauty of the island. One can ask a stranger for help and get it—willingly and gladly—because in reality no one is a stranger there. Kurt and Carol had raised dairy goats at one time and if a kid were born, he would know what to do until I returned. Many of the kids needed feeding with a stomach tube that season and I clambered over icy banks surrounding the barn with the apparatus in a pan of hot water to keep it from chilling the milk I would give the newborns. I had vowed that year to keep Gilda away from the billy, but she became pregnant along with the others. Donna, as well as Lee and Karen, who also have sheep, worked with me from eight-thirty at night until eleven to deliver her baby. At one point I called the vet and he advised, if all else failed, to cup a knife blade in my hand, go into the goat and try to sever the head of the baby from its body in order to get it out. We managed to work the infant out without having to resort to such dire measure, but by this time, the kid was dead. That night I raged mentally, venting my frustration at the lag in getting up a fence, but my anger subsided. There was no point holding onto it.

Later that spring Peter arranged to sell forty animals, all does and their recent offspring, as a startup herd. I had to let many of my favorites go and this was hard indeed because I cared about their welfare and felt I was betraying my silent promise to provide a good home to them for their lifetime. It was true that we were more or less at capacity and that in a year when Peter had not wanted to breed, we had had a generous crop of kids.

I understood these things but I wanted to protest that less than two years ago we were buying new goats; that I understood he had no time to put up a fence, but why not hire someone to do it? Peter came up for the sale and we entertained the couple who bought them with a celebratory dinner to commemorate the start of their new venture. I did not like the man, who seemed arrogant and rough as he chased goats to put in his truck. I pressed on the couple several pages of single-spaced typewritten instructions, showed them how to trim hooves and gave them extra medicine, wormer, and feed. At the last minute, after all the goats were loaded, Peter said that I did not have to let any go that I didn't want to. But it was too late for that. I tried to tell myself it would be a good home but felt depressed and guilty as Mimi and Tosca and Marguerite and others peered through the slats on the back of the truck as it pulled away. How did they fare? Not well, I think, from the occasional reports from the couple. The woman told me that two died after two months, from lice. But no, she couldn't tell me which ones because she didn't know. The last I heard of the goats, secondhand, was that the couple gave up the project.

From time to time Peter suggested we visit a commercial mohair-spinning operation I had discovered in Canada. The object was to look over their business with an eye to developing our own. But the proposed trip was always set vaguely for six months hence when Peter would not be so occupied with his business in Indiana. In the meantime, I told myself that the experience of raising the goats, being with them, was worth it, no matter what happened to our project. And indeed it was.

Elsa

It was earlier in the year in which we sold so many does and kids that Elsa became ill. She was one of our original twenty does from Texas, a pretty goat with horns that curved back evenly, flaring delicately at the tips. Her dense coat was well distributed, full around the neck and shoulders and reaching to her slender ankles. She wasn't a markedly heavy-shearing goat but produced a respectable amount of fine, curly mohair. She did not immediately distinguish herself as an extrovert; she did not seek my attention in the manner of Lucy and Mimi, Tosca and Carmen. But she responded quickly enough and became one of the friendlier goats from the beginning. Such a lovely gentle animal, I felt, should be called by no less a name than Elsa, for Lohengrin's bride.

She gave birth to a healthy male kid that first year, conveniently at ten-thirty in the morning; she needed only a little help from me, a slight tug to get the baby's shoulders through the opening. First-time mothers are hardly ever full-grown and are more likely than older goats to need some assistance, particularly if the baby is large. Elsa knew what to do and immediately began licking her baby dry, all the while uttering soft little mother-goat grunts. She had plenty of milk and was so solicitous of her baby that I had little to worry about. Her subsequent progress as a young mother and still-growing goat was uneventful. She never resisted having her hooves trimmed; she accepted worming and delousing graciously; I never had to use a shepherd's crook and acrobatics to catch her the way I did with Brunehilde, Gilda, and Louise.

She breezed through the trauma of separation from her kid when weaning time came, then in due course she shook her tail at the billy to signal her interest. As the winter wore on, she and the other does grew larger and more placid; in the last month of pregnancy, Elsa liked to stand quietly and let me scratch the base of her horns. Twins were born to her, one of which died. After losing the male of her twins, Elsa remained an attentive mother and in due course was separated from her vigorous daughter.

One cold, gray day in January I noticed Elsa shivering in the barn. Beneath her heavy coat she felt thin, and standing off to the side she appeared tentative in manner. Like a mother with children, goat keepers notice something amiss before it can be readily analyzed. A goat may just look odd: its tail may be down, or the goat may be standing away from the group, or its head may not be erect in its usual manner. There are any number of things that can trigger concern simply because they aren't consistent with the way a particular animal comports itself. When I suspect a problem, I whip out a thermometer and take the goat's temperature. If I'm lucky and there's no fever, I simply make a mental note to keep watch for further developments. If the temperature is above normal, I consult the vets and generally start medication.

Elsa's temperature was one hundred degrees, two degrees below normal. To further evaluate her, I offered her grain and was relieved that she ate it with gusto. I finished with a dose of wormer, injections of vitamin B and selenium, and returned her to the goat area. Next I called the vet.

"It could be serious," he said. "A subnormal temperature is as much cause for alarm as a fever. Start her on penicillin and warm her up."

I set up a kidding pen in the maternity room and rigged up a heating lap. A healthy doe won't tolerate a kidding pen unless she's occupied with her newborn; she'll jump out even if she's hugely pregnant. Elsa submitted to the pen and showed no interest in the heat lamp except to stand quietly under it. The next day her temperature had risen to one hundred three, slightly above normal. I moved her pen to the adjacent room where she could see other goats through the open top of the Dutch door. She immediately perked up, stretching her neck toward the others when they jumped up against the door to peer at her. But she now refused to eat grain, and in the following days her temperature hovered at one hundred four.

Penicillin had always before reduced a fever within a few days, but not with Elsa. After a week the doctors sent up an additional antibiotic. "Easier on the animal," they said. There was no change.

By January 31 Elsa had been sick for twelve days. I was discouraged and puzzled. There were no other overt symptoms, such as difficulty breathing or runny nose or diarrhea. That day, a Sunday, I had a surprise visit from one of the vets, who had flown up to the island to deliver a calf. He listened to Elsa's lungs and stomach, then gave her an injection to reduce the fever. "As long as she's eating hay, I wouldn't worry about her not eating grain," he said. Her thin little legs were getting sore from all the shots, so he said I could give her penicillin injections in the skin over her shoulders and the antibiotic Naxcel intramuscularly in her leg. He suggested aspirin, in the form of a veterinary bolus or large pill, which I crushed and mixed with molasses. I also started feeding her yogurt to maintain the bacterial mix in her rumen.

Her temperature seesawed. One day there was a fever, the next it was subnormal and I'd then put her back under a heating lamp. Sometimes it settled at normal for a day. As the days wore on, Elsa's illness lost some of its emotional intensity for me. She wasn't getting better, it seemed, but neither was she getting worse, and I had other concerns. The goats were due to give birth early that year, starting any time after the first of February. I hadn't planned it that way. In fact, it hadn't been planned at all, but that was when Peter had not been able to put in the fence in time to separate the does from the males. The shearers weren't scheduled to come until late February, so I was busy giving the necessary pre-kidding inoculations and shearing the does down the back legs and in the udder area to ensure that the newborns would not try to suck a lock of hair. I attended to Elsa every day, but not with the concentration I would have liked. Her condition stabilized, though, and I stopped taking her temperature daily. We changed medication once more. Her temperature suddenly plummeted and she began shivering, so I reintroduced penicillin along with Gentamycin, another antibiotic.

I hesitated to have Elsa shorn in her weakened condition. It was cold, and we were shearing earlier than I liked because it was the only date the shearer was free to make the long trip to the island. Ultimately, I decided to shear her to rid her of lice, rather than risk weakening her further. The best way to remove lice is to shear the animal and then use delouser. Strong, healthy goats are not plagued by lice as severely as weak ones are, which suggests that something in the chemical makeup of a well-nourished animal is less palatable to a louse. It seemed best to remove Elsa's coat despite the risk of exposing her to cold.

The kids started coming February 9; the temperature dropped below zero. Day ran into night for weeks. Babies came in the early hours of the morning and had to be whisked into a pen and under a heating lamp if they were to survive. Just as things settled down, the shearers came and we worked well into the night the first day and most of the second. It was still cold, colder than I anticipated for late February. Elsa was safely under a heating lamp, but the other goats, all pink-skinned and naked, huddled together in the barn, shivering in the cold. The day after the shearers left, I noticed one of the yearling wethers lying down near the hay feeder. His head was bent down as if he were studying the bedding of hay, and one of the old wethers we bought from Ingrid was pressed against him as if to warm him. I dragged him away and took his temperature. It was Elsa's first offspring; his temperature was subnormal. I put him under a heat lamp in a kidding pen, but he failed to show any response. I waited a few hours and when there was still no improvement, I covered him with an old quilt and climbed into the pen to massage him. Still he showed little sign of life; he wouldn't stand or eat or drink water. By night I was alarmed, and so with a neighbor's help I loaded him into a wheelbarrow and pushed him across the icy lawn, then hefted him up the stone steps at the front door and into the house. I settled him in the powder room with a quilt under him and warmed him with towels heated in the clothes dryer. I gave him dextrose and saline solution, as well as vitamin B and penicillin, and continued massaging him with the idea of helping circulation. That night I went to bed late, wondering if I'd find a dead goat in my bathroom the next morning. But the next day, when I opened the door, he was standing up, peering around and looking quite chipper. I led him into the larger bathroom and opened the window to accustom him to a cooler temperature. By noon, I felt I could take him back to the barn. That it was Elsa's son alone who suffered hypothermia is still a riddle.

By this time Elsa was in a more stable condition, with her temperature remaining normal. None of the other goats had become sick, so it seemed safe to keep her in a pen along with the mothers and newborns in the maternity room. And I was pleased to see that the activity there raised her spirits. She liked seeing the other goats and their babies, and would stretch her neck to peer over the side of her pen into the next. She never showed any of the aggressiveness some goats display to their neighbors in the maternity room, but instead simply seemed happy at having company.

"Why don't we take a blood sample and send it down to the state lab," Dr. Franks, one of the doctors at the Peninsula Veterinary Service, suggested over the telephone one day. "Then we can see what's really going on with her. You won't have any trouble drawing blood, will you?" he asked.

"What exactly do I do?" I asked.

"Just find the jugular and fill the vials I'll send up."

I urged him to give me explicit instructions, then practiced locating the vein a few times, pressing on it to make certain by the way the blood collected and caused it to stand out. Then I called a friend to come and hold Elsa while I drew the blood. It wasn't the sort of thing Priscilla would want to do very often, she admitted, but she held Elsa by the horns while I poked around. The vein wasn't so easy to find this time, partly because it was dark outside and the electric light on the ceiling was too high to be much use. I turned Elsa to one side and then the other, trying to get the angle right. Eventually the vein popped out and I drew a quantity of blood and squirted it into the vials. I sent samples of it for analysis. The results, in the end, were inconclusive: Elsa had an elevated white blood cell count but there was no hint of anything else amiss.

A month passed after the shearing. The goats were all growing hair and by then were blanketed with a thick, white stubble; some fleeces were even long enough to show curl. But little Elsa remained essentially naked except for wisps under her chin and a slight growth along her backbone. Angora goats grow hair at the sacrifice of their bodies, with nourishment put in the service of making fleece, while under their coats they may be literally skin and bone. Elsa never stopped eating, but her little body continued to lose weight. Apparently, she could not utilize the food she consumed; something impeded the absorption of vitamins, minerals, and calories. With almost all the goats delivered of their kids and the babies frolicking beside their mothers, I could turn my attention back to Elsa. I was determined that she get better, that somehow I'd find the perfect remedy, the right balance of nourishment and attention and will in order to halt the wasting that ravaged her body. I talked with goat raisers, with large-animal veterinarians, and with Ross, whose veterinary practice includes homeopathic and vibrational therapies. When I could bear pricking her no more, I ground vitamin tablets, mixed the powder with yogurt and molasses, and spooned the mixture into her mouth. She was fun to treat. She remained alert and eager and cultivated a delicate greediness. The friendly, amiable disposition she had shown early in her life blossomed into an impatience for my attention.

I constructed a large pen and put it in the room adjacent to the maternity room so that Elsa could benefit from the sun through its skylight. Spacious and clean with the light adding cheer, it pleased me. Every other day I confined Elsa to another pen while I cleaned hers, removing the old hay and spreading barn lime and fresh bedding. It gave me satisfaction: clean sheets for the invalid. I have no idea whether it affected Elsa the same way.

Elsa seemed to enjoy watching the yearlings in the adjoining section of the barn, craning her neck to see them better. So one day when the weather was particularly mild I decided to introduce her into the yearlings' area so that she could go outside with them. Putting her with other adult goats would have been too risky; goats tend to be abusive toward weaker peers, but the younger, smaller animals seemed safe enough. Besides, her daughter was among them, although I had no idea what kind of bond could be expected after months of separation. The daughter, a strong, beautiful animal, had tightly curled, shining hair that at the last shearing weighed the heaviest among the yearling fleeces. I don't know whether she had been aware of her mother next door, but as soon as I guided Elsa into the yearlings' barn, her daughter came up and stood near her. Other goats tried to butt Elsa, as goats do when there's a newcomer, but the daughter stood her ground between Elsa and the others and prevented any attacks.

At feeding time that evening, I sprinkled grain in the yearlings' trough and hoped that competition would whet Elsa's appetite for oats and corn. But there was no opportunity, for she was quickly knocked down in the boisterous rush to get to the feed. Righting her before she could be trampled, I realized it would be better to keep her in separate quarters for feeding and for the night. As I led her away, her daughter crowded next to us, refusing to leave her mother despite the dinner the other goats were gobbling. As a skittish yearling she was fearful of me, but this did not deter her from pinning herself to Elsa and being led to a strange pen.

There I fed the two goats and let them remain together for the night. At this time, I was trying all manner of food to help Elsa gain weight. She never regained her enthusiasm for grain, even when I mixed it with honey and brown sugar or with molasses and cooked apples. Sometimes I warmed it and sometimes I tried the more finely ground grain for kids, serving it up as a sweetened, warmed mash. None of these appealed to her, so to get oats into her, I tried cooking oat bran cereal and mixing it with raisins, apples, and molasses. Whenever I had a new idea, a new way to tempt her, my hope surged, then faltered when Elsa showed no interest. The oat bran, though, was modestly successful. My mood soared as I spooned it into her mouth and watched her tongue flick in and out to clean the sticky lumps of molasses smeared on her face. Elated, I gave her more. She lapped some from the bowl but soon grew bored with it, preferring the attention of being spoon-fed. Preparing Elsa's cereal in the morning became a pleasure; offering it to her was a joy mixed with hope and resolve for her recovery. When she apparently tired of oat bran after two days and refused it, I made oatmeal instead to tempt her. Seeing her welcoming and alert, eager, and, best of all, hungry, became a high point of my day. I think she knew it, for she became demanding in a lovable way, teaching me to vary her menu by rejecting any treats offered too many times in a row. But all this she did with an insouciant air, which utterly charmed me.

Once the oat side of the diet was solved, I still fretted

about how to get corn into her. A mash of corn meal was futile, as was canned corn; cooked frozen corn was better, but it was never a sure thing. Then I struck on tortilla chips made of corn meal. Elsa loved them. I'd crush them and offer them to her in a small bowl. She ate them eagerly, combining gusto with daintiness. She'd gobble a mouthful, then raise her head, look steadily at me and flick her tongue in and out as if savoring the piquant saltiness of the chips. Then she'd thrust her nose into the bowl for another ladylike mouthful. Even her daughter, who showed no interest in cereal, eagerly sampled the corn chips I offered her.

With the introduction of corn chips, frozen corn, and oat cereal, the carbohydrate and fat part of her diet seemed assured. I was still concerned about vitamins and thought a fresh source would be preferable to the vitamin mix I continued to force on Elsa. So I tempted her with raw vegetables— not merely leftover lettuce leaves and broccoli stems, but fresh asparagus, parsley, and spinach, which I'd sort through to find the choicest morsels. By mid-April the grass had begun to green on the lawn and enough grew on the protected side of the barn and house for me to pull handfuls to offer to Elsa. I even combed the woods, where longer clumps of grass grew abundantly.

Elsa's daughter remained wary of me, but never did she falter in her consistent devotion to her mother. After their breakfast on sunny, mild days, I released them into the yearlings' pen, which had access to its own pasture. Much of the time Elsa would settle herself on the ground in a sunny spot that was sheltered from the wind by a wooden fence, her daughter standing guard near her. As the days warmed, the rest of the herd moved farther away from the barn to the distant end of the pasture. One day, when the yearlings congregated at the far end of the pasture munching the still-brown stubble, I saw Elsa and her daughter slowly making their way along the fence line to join the others. They looked like a skinny, balding little old lady and her robust companion. The daughter never outpaced her mother by so much as a step, but rather kept at Elsa's side, perfectly matching her own progress to the slow pace of the weaker animal. I wondered then if I would have been as good a daughter had my mother lived to be old and frail. Such a devoted daughter deserved a worthy name: she became Ruth to Elsa's Naomi.

Early May is still cold on the island. Birches begin to leaf only after the second week; lilacs often don't bloom until June. I worried about hairless Elsa chilling in the fresh spring wind, but she gave all indications of liking her schedule of being allowed onto the pasture during the day and residing in her pen with Ruth at night. I pondered her illness. There are so-called wasting diseases in goats, but my books were not specific about them. Her condition sounded to me like Johnes' disease, a fatal illness in which the animal weakens and becomes emaciated over six months' time and finally, in goats, develops diarrhea at the very end. The veterinarians thought it unlikely; for that reason and because results of testing take a minimum of fourteen weeks, we never tested a fecal sample for that particular illness. Whatever was wrong, it seemed clear that something had destroyed Elsa's ability to utilize nutrients. Aware of the course of Johnes', I noted with sadness when diarrhea struck on May 6. Elsa, whose stools had always been normal, suddenly expelled a massive pool of liquid feces. I cleaned up the mess and led her back into her own pen, where I cleaned her as well as I could.

I returned to the barn for Ruth and noticed for the first time that she was stiff-legged and listing to one side. I dragged her into the pen and took her temperature. At one hundred six it was higher than Elsa's had ever been. I started her on penicillin and gave her an injection of Dexamethasone to reduce the fever. My friend Rachel, who had been helping me clean the barn and trim hooves, and I watched helplessly as Ruth staggered and drooled, an ear and lip drooping on one side and uneaten hay sticking out of her mouth.

Their symptoms were so different that I remain confident that Elsa did not infect Ruth. The yearling's high fever could have accounted for the staggering and stiffness. But these symptoms—the listing to one side and the very high temperature—indicate the encephalitic form of listeriosis, which is caused by bacteria. By the next day Ruth could not get up, although I gave her large doses of penicillin and doses of thiamin. Soon Elsa was lying on her side, too, but still more alert than her daughter. Before I went to bed on May 7, I checked the barn and saw that mother and daughter were lying in the hay, about two feet apart and seemingly unable to move. However, the next morning they were so close they were touching, although they were in the same posture as the night before. Rachel came by several times that day and we deliberated. Neither Ruth nor Elsa showed any sign of improvement, with Ruth now apparently comatose. I discussed with the veterinary office the advisability of getting the goats in for a postmortem, as it seemed apparent they would die. I wrestled with it. They would have to be kept on ice. There were not many boats a day to the mainland, and I was needed at the farm because several does were still waiting to give birth. I looked for Styrofoam ice chests at the store but none was large enough to accommodate a full-grown goat. I ruled out asking one of the commercial fishermen to lend me one of their ice chests for transporting a diseased, dead animal. I knew also that the autopsy the local veterinarians were equipped to perform was cursory, and that a detailed one at the University of Wisconsin vet school in Madison could cost as much as several hundred dollars plus a ten-hour round trip. With no other goats falling ill, a postmortem would have done little besides satisfy my curiosity.

Still undecided about an autopsy, on the morning of May 9 I began digging a large grave. Rachel was a loyal companion in the sad business of watching Elsa and Ruth die, and helped me dig the grave. At seven that evening, I saw that Ruth was no longer breathing. Her little mother seemed unaware for now she, too, was apparently comatose. Rachel came over with trillium from her woods to place on Ruth and we buried her, keeping the grave protected but open to receive Elsa. She remained unconscious, although from time to time I managed to give her water, which she swallowed without reacting to anything else. Perhaps because her little starved body was accustomed to existing in a state of deprivation, she managed to live for several more days. Early on the morning of May 13 I saw that she was dead. Rachel and I buried her in the grave with Ruth, sprinkling over her fresh trillium and daffodils and dried lavender from the previous summer's garden.

Alone

WITH THE FIRST cold winds of fall the landscape empties. It's as if anything not nailed down, not fixed firmly in place, will be swept away by the force that extinguishes autumn's colors and leaves desolation in its wake. But to islanders this seasonal cleansing allows room to breathe. The land once again mimics the vast expanse of water to the north and east, once again it's uncluttered, free of the extraneous. It's nap time for the land: gray covers drawn up, birds silent, summer's visitors gone. It's time for the regaining of perspective that a good rest provides. By September I crave solitude, but summer persists. Besides, there's shearing to be done and visits from friends. By October I think I might achieve it, but this happens seldom. More often Peter decides on a building project, which brings him to the island on weekends, often with friends. So time is taken cleaning the house, shopping and cooking, cutting the grass, and in general attempting to have everything suitably shipshape. These are fun times, too, but my life does not yet seem my own.

But November's desolation and unpredictable winds whittle visits to a minimum. It's a welcome time. People often ask me if I am not lonely living by myself on the farm. It's a question I'm usually asked by people who are visiting the island for the first time and I always suppose they are asking, as well, about the larger context of living cut off by water from other towns and cities. What summer visitors don't see is the pulling together in the winter of the island's community, which must be very much like farming communities of years ago when, once the harvesting was done and the farm made ready for the ice and snow, people had time for social life. Although I do not belong to a church here, every October I am asked to provide four quarts of acorn squash for the annual harvest dinner held by the Lutheran Church. Every year I buy acorn squash, wondering how many I should cook to produce four quarts. And I marvel at the blatantly old-fashioned quality of the request. Thirty years after Betty Friedan's *The Feminine Mystique* I'm still being asked to do something that would not be asked of a man in a similar situation. And I'm happy to comply. I could be busy all week with the various women's organizations or informal Bible study groups or choir and singing groups or service clubs. But I choose not to, hoping I am not offending anyone. At the same time I know I am denying myself a rich slice of community that I'll not have another opportunity to savor.

Somewhat removed from the island's social life by my own choice, I am never lonely at the farm. My world as I write this, somewhat after four o'clock on a Thursday morning, is busy with personalities—my dogs and cats—making themselves known despite the early hour. Some time ago I realized that the only way to be sure of having time to write was to get up at three-thirty in order to be at the word processor by four. My energy level is higher in the morning for most pursuits, whether it's writing or goat chores or practicing the piano. For months, years even, I told myself mentally that once I cleared away whatever had to be done with the goats—hoof trimming, barn cleaning, moving bales of hay one at a time by plastic toboggan over deep snow from the storage barn to the goat barn—I'd be free to write. That time hardly ever came. I'd finish cleaning the larger of the two sections of the goat barn, a monthly job that takes from sixteen to twenty hours spread over several days, and realize it would soon be time for worming the goats. Now I write first, then get to the chores.

Marcel, my aging Abyssinian cat, sleeps on a bath towel folded near my word processor. He purrs from time to time and wakes to butt his head against my hand and threaten to walk across the keyboard. Lily dozes at my feet, one paw resting on my shoe. It's too early for her to get up, she knows, but she doesn't like letting me out of her sight. Molly, my Staffordshire terrier mix, is in the living room in a chair by the wood stove, barking occasionally into the blackness, seeing or sensing something I cannot: a raccoon in the compost heap, perhaps, or a deer on the frozen alfalfa field. Only Francie, an elderly whippet whose delicate skin and thin hair make her seek warmth, stays burrowed under the covers of the bed I've just left. Teddy and Diego, the barn cats, are probably at the front door; when I let them in, they'll settle on my work table or on the carpet near my chair.

I never feel lonely with these personalities intruding themselves into my consciousness, nor do I feel I have the quality time with myself that I crave. That time comes most often on walks and on weekend mornings when I take a thermos of coffee or tea back to bed where I read, letting my thoughts build on the passages of the book on my lap. The dogs and cats settle themselves around me, sleeping and generally being considerate until I encourage them to stir.

Around seven this morning I'll feed the goats and walk Molly and Lily separately on leashes. Walked together, they'd arm wrestle, as Rachel describes it, while I stand around becoming chilled. Yet they must be on leashes because we are close to a road and both chase cars. They do not chase every car; Molly prefers trucks pulling boats, which she must perceive as a particular threat. Her chasing, when she has the opportunity, is fierce; she'd like to tear the vehicle apart. One April at the end of kidding when I had been too busy to give the dogs much attention, Molly charged a camper from the front and was hit in a way that damaged the nerves of her right front leg. The ferry schedule had just increased from one to four daily boats and I was able to take her across to the veterinarian in Sister Bay. Ultimately she lost her leg, but this has not deterred her from chasing. Perhaps it's only strengthened her resolve to rid the island of vehicles. Lily will chase the occasional car, too, but for her it's merely a race, something she delights in. I've used a borrowed shock collar to break the habit but the lesson receded from her brain after several months, so now I simply make sure the dogs are leashed until we can get to open fields. After our outings, I'll return to the goat barn for the work I've planned for the day.

This week I've been trimming hooves of the males and cutting away their urine-soaked belly hair to prevent pizzle rot, which is the colorful name for the chronic condition called posthitis. The disease is attributed to a combination of protein in feed and the bacterium *Corynebacterium renale*. The goats excrete urea in the urine, presumably a higher percentage of urea when the intake of protein is elevated, and this is converted to ammonia by the bacteria that live in the area of the male goat's penis. The ammonia ulcerates the skin and causes scabs and scar tissue. I first noticed pizzle rot when trimming hooves. At that time the goats who suffered from it must have had it for some time, because it looked as if their penises were covered by a granular mud cake that broke apart when I applied warm, wet cloths to the area. Treatment involves peeling away layers of oozing, dying skin, cutting away hair, washing the area with disinfectant, and spreading a broad-spectrum antibiotic cream over the affected part.

It took some research to find a name and description of the condition. Dairy goat books do not name it, probably because male goats that grow to maturity are the few maintained for breeding. Likewise, most ram lambs are sent off to the slaughterhouse. I suppose if there were a greater taste for *cabrito*, or kid goat meat, not many male Angora goats would be kept. A doe that provides a kid or two every year in addition to her fleece is more economical to have than a wether. But we keep all our goats, and pizzle rot is a constant problem. I fiddle with the feed, adjusting the males' ration of corn, oats and soybean mixture for protein, but it makes no difference. Our hay doesn't have a high protein content either and I suspect the condition is a function of the damp climate. Indeed, pizzle rot is more noticeable in the winter months. Preventive measures help most of all, and these involve trimming all the hair away from the penis area on three separate occasions between shearings. It's vigorous work. I must catch each animal and drag him outside his pen so that I will be able to concentrate on clipping around a sensitive area without being harassed by the other males, who like to nudge me out of curiosity and butt my patient out of sheer goatliness. To get the rather large animal into position, I trip him by pulling his far hind leg and toppling him. Then I straddle him, pinning

his legs if he kicks, or if he's inclined to bite, keeping my boot-
ed leg over his face. I talk to the goats as I carry out these
operations, to calm them and to encourage myself. Some are
cooperative, others are visibly upset and tremble with fear. I'm
sure that when the pizzle rot is advanced, the procedure is
painful, but goats are rather stoic and my resolve is firm.
Usually I give them a handful of grain afterwards, especially
the fearful goats, because I hope that in some recess of their
goat brains they will associate the experience with something
not totally unpleasant. I work slowly and carefully to avoid
cutting them with the new pair of haircutting scissors I buy
for the job. To do all the wethers, both adults and kids, and
the billies takes several days, and I generally come away with
bruises.

Occasionally Peter has helped me with this chore. He's
both firm and gentle with the goats and the best assistance I've
had at holding the goats when I trim them. I've learned to do
this kind of work for only a few hours at a time; I've learned
that I don't have to complete a chore in one day, that nothing
will happen if I don't forge to the finish except that I'll be
much fresher the next day. It's cold in the barn these days and
raw, and the job is smelly, and for these reasons it's especially
pleasant to come in after a few hours and clean up. There are
always other chores to be done, things that have gone wrong
and need fixing. The billy may have broken through the fence
again; the door separating the male kids from the females may
have opened and the goats will need sorting into their respective pens; a gate may have worked off its hinges. At some point
in the day, I'll drive to the post office for the mail and pick up
anything I need, such as more screws or lag bolts or new
hinges to replace the ones the male goats recently ripped off
the gate to their pasture. Midafternoon I'll walk the dogs

again, taking a flashlight with me for my four-mile walk with Lily because it will be dark by the time I finish. Later I'll check the goats and feed them grain and their evening hay.

When I've finished with the hooves and pizzle rot of the adult males, I'll trim the yearling wethers and male kids. Then it will be time to trim the hooves of the females. In between, I'll clean barns and move some bales of hay.

These days I haven't been working completely alone. Rick, a young man who helped my brother build a new hay and implement barn and has helped me unload wagons of hay, is here making pallets. The new hay barn does not have a floor, so before hay can be stored in it, we must have something to put the hay on to keep it from drawing moisture from the ground and ultimately molding. A thick wooden floor would suffice; a cement floor would not. I will put down a layer of a thick grade of plastic and then the pallets that Rick is building from treated wood. Rick and I are not working in the same part of the barn, but I can hear his saw and feel his presence, and the companionship makes the work lighter and go faster. It's nice to know, too, that if there's a problem that my experience or strength is not equal to solving—if a billy suddenly turns belligerent or I'm stumped by a repair job—I can call on Rick to help.

I feel the same pleasure when Peter is here. We hardly ever work together at the farm. He has his own projects; I have the routine chores of goat maintenance. But there's still companionship. We convene at noon to go out to lunch. In the evening, I change to a skirt and sweater while Peter puts on a white shirt and good trousers. We have a Scotch first, then a leisurely dinner in the dining room with candlelight, cloth napkins, and wine. It feels very civilized compared with my normal evening routine, and the companionship, better-than-usual dress, and candlelit meal give the sense of being rewarded for the day's work. Then I wonder why I don't do this for myself when I'm alone. But something would be missing. Cloth napkins and candlelight are better shared.

February

Last week the daytime temperature slipped to minus ten and continued dropping through the night until the cold registered minus twenty. Yet there's a sense of impending spring about February's frozen island. There's a new lilt to the sound of the birds, which didn't chirp much at all last month, and the days are longer by an hour at each end, making two more hours a day to warm the earth. I can feed the goats at five in the afternoon and finish to see the last streaks of pink and orange in the western sky. My mood lifts with the awakening of the earth. Subzero temperatures in February seem a transitory thing, not like below zero in dark December when we can expect two and a half months or more of deep cold.

Afternoons I cross-country ski with Lily in the woods on a logging trail at the north side of the island. Along the forested shore, which can be seen through the trees from the trail, the ice is startlingly blue-green. Huge chunks have piled high close to the land, driven by the north wind. Stretching out beyond, white and lunarlike in its vastness, the ice appears rough but essentially flat until it meets dark blue water. Beyond the narrow band of cerulean blue a second line of dazzling white delineates the horizon. This is the ice that surrounds Michigan's shoreline and the islands that reach down from Michigan's Upper Peninsula as if trying to reconnect with their Wisconsin sisters.

The ice is surprising to me, and mysterious. It makes me think of Greenland and Finland, places I've never visited except in books. Each winter is different in texture, in cold. Some years fantastic ice sculptures—many-tiered caves fringed with gleaming icicles like stalactites, silent cities of frozen edifices designed by a wizard architect gone berserk— transform the western shore. These structures are fashioned from waves and spray freezing in midair. Last year there was no blue open water between Michigan's Garden Peninsula and Wisconsin; the ice was three feet thick all the way to the Upper Michigan shore without even a narrow channel of water. This year is milder, and open water allows room for the ice to shift with the wind, to pile chunk upon chunk against the shore until towering piles gleam blue and green in silence and isolation. This mystery and haunting beauty must be partly what binds people to the island once they've spent a few winters here—this and the sheer ego gratification of knowing they can meet the challenge posed by extreme cold.

At this time of year ice shanties or ice-fishing huts dot the frozen harbors on the south and west sides of the island. They're usually made of aluminum, lightweight so they can easily be pulled down the road on runners or loaded onto the back of a truck, and just large enough for a couple of men to sit in while they fish through ice at least two feet thick. It's a winter pastime for men, it seems, but seldom for women. Inside the shanties fishermen make themselves comfortable with kerosene heaters and cans of beer and wait for perch and lawyers—the local name for burbot—to nibble at a baited hook at the end of a line dropped through a hole drilled in the ice. The first time I went out to John's shanty, I thrilled at the sensation of riding over the ice. It was a freeing feeling; I felt my world expanded and as limitless as the ice we soared over, stretching out in all directions. A certain titillation of fear heightened the pleasure but was counterbalanced by the sight of other shanties in the harbor, other trucks parked near them. Great pools of black spread out before us. Were they safe? Was it softening ice? Apparently it was not. However, a short time earlier a man had drowned at Dykesville, south of Sturgeon Bay, when his van went through the ice. His son and grandson survived. I mentioned it to John.

"I don't think about things like that when I'm on the ice," he replied.

I was silent. I couldn't help pondering my escape from the truck just in case the ice started to give way.

Inside his shanty John handed me a cup of tea from his thermos and opened a bottle of beer for himself. Through the hole in the ice I could see green water and the occasional small fish swimming by, ignoring our lines. The ice didn't appear two feet thick but John assured me it was. The kerosene heater soon warmed us; outside I could hear the wind howling and I knew it was bitter. The fish never bit that evening and soon our talk shifted from fishing to goats. After an hour, John must have felt the pull of chores at home and we left, again sailing over the frozen lake. In Russia and Alaska certain rivers become busy thoroughfares in winter, the ice frozen so deeply and solidly that heavy trucks laden with tons of materials are commonplace. Here, the ice is deemed unsafe after March 15. After that time, anyone leaving a shanty on the ice overnight risks a fine from the state Department of Natural Resources.

Men's Day comes the third week in February, followed by the weekend Fishing Derby. On that Thursday, the men who care to do so declare a holiday, a day of letting down, revving up, and cutting loose from whatever constrains them, including, most likely, the ice and interminable cold. But it's celebrated by playing winter's own game, out on the frozen lake, snowmobiles racing across ice and snow. It's a cooperative affair, with everyone pitching in, supplying food or drink, cooking or tinkering with snowmobiles that refuse to start.

The festivities start with breakfast at the home of one of the men, the road to his house lined with snowmobiles. It used to be held at Peter's farm when our friend Jay lived at the house, before I came to the island. Forty men would crowd into the kitchen to eat eggs and sausage, drink Bloody Marys and screwdrivers and beer, and play a few hands of cards. After breakfast they point their mechanical steeds east and head for the lake and beyond, across the frozen stretch to Rock Island, whooping, hollering, and feeling no pain. They charge around Rock Island, cowboys of the frozen lake, and return to a steak fry at a house on the shore, then back to town, many to the taverns.

This year while walking Molly on Michigan Road, which runs west from the east shore, I heard snowmobiles and looked behind me to see a flotilla of the beetlelike machines, lights on, approaching in the late afternoon. Molly and I stepped aside and waved. They zoomed by, making a wide berth into the fields as they passed us.

The next day, Fishing Derby weekend starts. Thursday's palpable excitement extends through the next days as the ferry disgorges off-island cars during its extra runs Friday, Saturday and Sunday. It's a time for people, men in particular, to return to the island. The grocery store stocks for an overflow of visitors; the taverns, whose kitchens are frequently closed down for the winter, heat up their grills and put in extra liquor and food. People not seen on the island since summer break away from their winter routine in the cities to visit. But mainly it's three days devoted to fishing on the lake through holes in the ice. Snowmobiles and trucks and four-wheel drives dot the white expanse along with the shanties themselves. At the end of the derby, the prize catches—northern pike, perch, lawyers, brown trout—hang frozen, stiff as boards, from the porch at Nelsen's Hall.

February weather is as variable as any other on the island, but the extremes are at the cold end of the continuum. After a subzero plunge this year, the mercury is rising; the goat barn temperature registers close to forty degrees. It is probably thirty-five outside, warm enough to melt some of the snow.

The next day all surfaces freeze again into a slick lake covering lawn, drive and roadsides. The ground is far too unremittingly solid to admit any absorption of moisture, so the snowmelt simply sat on top until the warming spell passed. With the resumption of cold, an icy wind blows fiercely from the northwest.

I moved hay yesterday in the milder weather over newly fallen snow, working for two hours as I whittled away at the stack in the hay barn. I planned to clean the boys' side of the barn, which is overdue. It's easy put it off in anticipation of reasonable weather, because the goats must be shut out to do it. They don't like to be out in snow or rain and do not fare well when wet. Their door to the pasture poses the usual winter problem; it's blocked by a buildup of mud, manure, and ice, and can't be closed. To solve the dilemma I nail panels from kidding pens across the entrances. Then I further block the opening with our large, heavy hay feeders turned on end. The does accept this arrangement; the billies and wethers resist. One animal or another will stand at the barrier, banging it with his head, tearing at it with his horns. I work for four hours with manure fork and wheelbarrow, and progress is slow, partly because I must walk gingerly over ice and crusted snow to dump the waste on a huge mound at the east side of the barn. In the summer, friends who want it for their gardens will haul it away. Once I've raked the dirt floor of the barn as clean as possible I spread barn lime and over that I strew fresh hay. Then I dismantle the barriers until the next day, when I put them up again to finish the job. It pleases me to do these chores although they loom large before I start them. The exercise is good and sometimes I wear a portable tape player and listen to music or recorded books. But the pleasure comes from knowing that the goats' house is in order once again, clean and sweet smelling. And it's satisfying to see the animals push in when it's done, then stand around munching fresh hay off the clean floor.

Waiting for Spring, Fourth Year

It's not spring yet. The goat barn is still a chill twenty degrees at six-thirty in the morning, and the thermometer at the grocery store read twenty-two at noon yesterday. But spring is in the air. The days are longer, the sun higher. Yesterday I walked Lily after six, having spent most of the afternoon cleaning the barn. The night air was moist and sweet; the fields spread out as brown as ever, but underneath the dry thatch an occasional blade of green appeared. Wild strawberry leaves remain green all winter, protected by dried grasses, Queen Anne's lace, spotted knapweed, and a thick layer of snow. Underfoot, the ground's now spongy in places and squishes where the sun has persuaded its moisture to thaw.

Last Sunday the temperature climbed well above freezing and snow receded from the fields, leaving gray-brown swaths. Deep snow remained only at the perimeter of the pasture and in high banks around the barns and house. With the imminent thawing it seemed time to loosen the fence wires so the corner posts wouldn't be pulled from their foundations by the shifting of the thawing ground. It should have been an easy chore, particularly in the light wind that was no longer sharp that day, but I had not anticipated the depth of snow as I tramped the fence line looking for the ratchets on the seven wires. I lumbered slowly through knee-deep snow heavy with moisture, now too soft to support me on its crusty surface. I thought of returning to the house for snowshoes. Too cumbersome, I decided. Woodpeckers drilled in the trees, and I heard the distinctive, almost liquid honking of a sandhill crane. It was tiring to make my way, and when I got to the far fence, the four bottom wires were still buried in hard-packed, icy snow. I dug out one ratchet with a small spade and loosened the wire, but it was so stiff and unforgiving that I decided to leave the other three until the snow melted a bit more. It's only March, after all, and we will surely have more snow and cold before the ground truly thaws.

It's been a cold winter. The temperatures dipped to thirty below zero one night, and fifteen to twenty below is usual. We're prepared for it here. Some people have had to thaw their pipes but it's not like other parts of the country when hit by a deep freeze. A good friend, Dick, who had been the editor and co-publisher of the newspaper I worked for in Santa Fe, said that once the temperature in Santa Fe dropped to forty-seven degrees below zero; natural gas stopped flowing and most homes were without heat. Here, I bundle in layers and postpone any chores that necessitate gloveless hands, at least until it's twenty degrees in the barn. I can't complain about this winter with its vivid blue sky, but I long to throw off my coat and run freely and easily, unhampered by tall heavy boots and constricting layers.

Here, surrounded by the cold, deep waters of Lake Michigan, the land lies locked in ice, chilled to temperatures below those normal for March elsewhere. Densely packed ice breaks up, drifts apart, and then is blown together again in tight formations that hang onto winter's cold. The phrase "icy grip of winter" has new meaning. When other parts of the state are treated to occasional mild days—say in the sixties or even seventies—to give hope and a taste of the season to come, on the island we're lucky if the mercury reaches forty. We wear our ski parkas into May.

This year talk was of the Coast Guard's huge icebreaker, the *Mackinaw*, that is to be retired this spring after some fifty years of ramming through ice in March to clear a shipping channel so the great ore boats and other cargo vessels can resume traveling the Great Lakes. The 290-foot *Mackinaw* can break ice that is between four and six feet thick, but this year it was stranded in the harbor at Escanaba, Michigan, daunted by ice too thick to break through. The enormous icebreaker sweeps, or rather crunches, down the middle of the lake, just passing Rock Island. John said that this year when it passed, he and Patty stood in front of his house, almost at dead center of Washington Island, and heard it. This year everyone talks about the *Mackinaw* because they're concerned the Coast Guard will replace it with a boat less than half its size. The *Mackinaw* may be too old to maintain, but a smaller boat won't be able to do the job the *Mackinaw* does.

This year I'm noticing spring's arrival, noting the subtleties that I've missed the last few years. My first year here I didn't know how to look for spring until May, when the birches begin to leaf. New to everything, I waited in vain for sweeping changes in March and April, and was disappointed that the landscape persisted as unrelieved brown stretching to the fringe of gray at the horizon. Temperatures seemed little different from winter, close to zero at night and just above freezing during the day, melting the snow slowly, causing it to recede and leave in its place dried stubble in the fields, and a thatch of lifeless grass where the lawn once was. It seemed that winter ended sometime between May 15 and Memorial Day.

The next three years the kidding absorbed my attention and thoughts, my imagination, even my dreams, soaking them up like blotting paper. The days of March and April were simply numbers on a calendar without content and context, mere hooks from which to suspend the births and occasional deaths of kids. Kidding, which lasts about three weeks in peak intensity although it may stretch out longer, is a hallucinatory time. I remember it the way I recalled childhood illness, when a high fever caused me to drift in and out of sleep and consciousness. Days and nights run together during kidding, daylight jumbles with night and becomes confused. At two in the morning I walk out into the frigid air, feeling only the freshness, not the cold, and I wonder if it could be ten AM, say, and I simply hadn't noticed. The fact that it's light or dark seems of no importance, makes no difference. This year kidding comes later. And because I'm more experienced, perhaps it won't demand all my waking and dreaming thoughts. In the meantime, I'm noticing spring.

Spring on this frozen patch of land reveals itself in the light as it lengthens at either end of the day. Thick and rich, it now comes from a sun higher in the sky and has a golden honeyed quality to it compared with the brittle, thin light of winter. It doesn't warm too much yet, but its promise of warmth is genuine. Warmth will come with the waiting. And even if I am unready to shed my many layers just yet, the sun touches us in other ways. The snow now appears scabrous, patchy, a damp crust on top of stubble with great hollows underneath. I imagine a whole world engaged in busy, coming-to-life-again activities—insects, grubs, voles, and other small rodents stirring languidly first, then quickening as the days lengthen and they prepare for the final burst of spring. The goats, too, feel the sun now and lie on the snow at the south side of the barn where they're sheltered from the sharp north wind. The air feels softer some of the time, and when the south wind blows, it's not so heavy with rawness.

In March, bird song embellishes the lacy tangle of bare branches near the house. At dawn a mourning dove—my father used to call them rain crows—sings its yearning. I hear a distant owl, and later when I walk through the woods I hear cardinals, but never see one. John says to look at the top of the tree. I keep looking but see only a pileated woodpecker, roosterlike with its large body and red head. Near the first of April, clouds of robins descend. This morning hundreds, it seems, are hopping about on the winter ground. Later they're in the apple tree, pecking at the withered, thawing apples that hung all winter like forgotten Christmas ornaments.

Two days ago spring retreated. It snowed all day until many inches accumulated. It didn't melt away as it would in most places, but hardened under the north wind's blast. In hours I was plunged into winter as I clambered over icy ridges around the barn. One winter I nailed plywood sheets to the ice mounded in a steep bank around the barn so that I could enter without sliding backwards. From the barn a three-foot-long icicle hangs in the company of shorter ones. These and the frozen ground underfoot make it feel like the dead of winter but, still, the light tells me differently.

Three days pass and the snow again recedes, the wind

shifts to the south, softer but now damp as well as cold. I fret because it blows into the goat barn where the newly shorn animals huddle together, shivery in their sleekly silvery-pink skin. I walk against the wind with my dogs and try to see the beauty of the monochromatic winter landscape, all pigment pulled from the trees and fields by the departed sun, as if a drain had been unstopped and the color had slipped away. The fields appear soft, although I know they bristle with dry stubble. The gray blur of leafless trees looks soft, too, and I know that the beeches and birch are beginning to swell at the tips with new buds, adding to the effect.

Hay

EVERY SUMMER the backdrop to my days is hay. For several years I wrestled with the same questions: how much to put up, where to find it, when to cut it, whom to get to do the haying. I still calculate the amount of hay the goats will consume from October through May when there's no pasture for them, but now I buy most of our hay off the island, which eliminates some of the uncertainty. The first years, though, getting in hay meant finding fields to cut and a farmer with a mower and hay baler to do the work. I learned I could do only so much to move this process along, dependent as it was on the weather and the whims and commitments of other people. But still I worried, as if the input of emotional energy would help.

Lacking good pastures on the farm and a diversity of the woody cellulose-rich vegetation goats prefer, we give our herd hay—ideally composed of alfalfa—as the staple of their diet in all but the few months of summer. Then, or at least when there's ample rainfall to encourage the pasture, the goats graze the thirty-acre enclosure that makes up in size what it lacks in lushness. In addition to the pasture, my brother has one sandy five-acre field planted to alfalfa, which is only a small portion of his 145 acres. Most of his acreage remains uncultivated, waiting for the year when he will finally have time and equipment to prepare the soil and plant new fields. He reads constantly about novel and still-experimental ways to improve the soil, and attends workshops and conferences whenever he has the opportunity in anticipation of the day he can put his theories to work.

The problem with growing hay on the island is the short growing season. Many farms in the Midwest average three cuttings of alfalfa hay a summer; on the island we seldom have even two, partly because of the weather and partly because of the difficulty getting someone to cut and bale. Just six miles across the waters of Death's Door the land warms faster in the spring; grass is a rich green by the end of April, daffodils bloom, and crops push up vigorously, washing the landscape with shades of green. But surrounded by ice sometimes through April and into May, the island takes its time to thaw. Here tulips open well into May, lilacs bloom in June, horsechestnut trees in mid-July. The first crop of the year's alfalfa, harvested in June elsewhere, does not show its purple flowers until as late as mid-July. The delay is a function of cold weather and frequent dry spells during April, May, and June, when farmers elsewhere normally count on spring rain. When the crop is finally ready to harvest, when the first flowers open, the July rains start. The maxim "make hay while the sun shines" took on new meaning that first year with the goats. Summer showers made it a trick to get in the hay, because haying requires three consecutive sunny days: one for cutting, a second for drying in the fields, and a third for raking and finally baling.

Getting someone to cut and bale the hay is another problem. Our operation is too small to warrant spending thousands of dollars on a tractor to pull a mower or baler, which themselves are costly even when bought used. The island once supported many farms, particularly dairy farms that raised crops to feed livestock. But those days are long past and only a few farmers remain who have equipment to make hay. Those farmers work part-time at it; they're contractors or commercial fishermen full time and fit in farming when they can. So we're dependent on people whose first priority is their own work, then their own crops and those for a few longtime customers. Too often our hay was harvested so late a second cutting was not possible; another cutting would not allow time for the plant to come back enough to survive the cold winter.

The first year, when we had only twenty-one goats, any extra hay we needed we easily bought from local farmers. That year John, who had arranged for planting and harvesting Peter's alfalfa field several years earlier, took charge. He arranged to have the field cut and he himself baled it. A hay baler sweeps down windrows of cut hay, scoops up a pile, compacts it into tight bales, binding them with twine, cutting the twine, and sending each bale through a chute that overhangs a hay wagon. A farmhand balances on the wagon as it jolts over rough fields, pulling off each bale and stacking it. I was that farmhand for John, although the very first time I did it I had little idea if I had the strength or the balance.

The day was hot and humid but I dutifully followed suggestions and wore long sleeves and jeans because alfalfa will scratch, and any exposed skin becomes a crisscross of raised red welts. The bales chugged up the chute slowly enough for me to get into position to grasp each one. Teetering on the wagon, I took hold of a bale by the twine and tugged, trying to swing it in front of me toward the back of the wagon, but it thudded to my feet. I lugged it across the platform, then resumed my stance for the next one while gritting my teeth and resolving to do the best I could.

After a while John asked if the bales were all right—not too heavy. How heavy should they be when first baled? Surely heavier than when they'd dried a bit, but I had no clear idea. I was eager to do my share of the work, however, so I said they were fine. John must have noticed my struggles because a short time later he stopped the tractor, got off and lifted one of the bales himself. "They're heavy," he remarked. "We'll let them dry another day." When we resumed a couple of days later, they were much lighter and I could easily pull them off and stack them. Still, it was hard work; determination made up for lack of strength until the strength developed. Perspiration ran down my face and back and cooled me when a breeze picked up. With a hoist aided by my knee I managed to stack bales on the fourth tier but no higher, so John stopped occasionally to fling bales five and six tiers high, making more room for me to work on the wagon. Once we had filled two or three wagons, we covered them with tarps or if there was time, unloaded them into the barn using an electric conveyor. It was fun to bale with John those first years; he was considerate and went slowly enough so that I could manage the weight of the bales on a jostling platform. Later, I tried to bale for Martin—a local builder who also farms—when he brought in our hay, but could not. He drove the tractor so fast that it was all I could do to remain upright. It took two people with Martin: one maneuvering the bales to the stack and the other stacking them and keeping the listing rows in place.

That first year John and I baled my brother's field and one of Martin's that he cut for us. Then we baled Peter's field a second time. The second cutting, or second crop of alfalfa, as it is known, is more nutritious than the first and I was pleased we could get it in. John and I worked late in the day after he finished at his building jobs. We'd work hard baling in the heat of the late summer afternoon, then exhausted and thirsty we'd drink a can of beer and share the mood. Looking out over the fields, the goats cropping the pasture in the deepening twilight, the whippoorwills beginning their nightly refrain, it was a time of contentment and of enjoying the fullness of the moment.

That second crop of hay turned out to be too wet, however. I had begun following the weather reports although they seldom applied very accurately to the island. Rain was predicted and even though our hay was a little too wet to bale, it seemed better to chance it than to have it thoroughly drenched and ruined. When it was loaded in the barn, John instructed me to check it several times each day by inserting my hand between the flakes of alfalfa. If it was too hot to bear, I was to call him at once and begin hauling the bales outside. When a barn burns, he explained, it's often because hay was baled too wet and ignited from the heat generated by drying. I checked for several days until the danger had passed, but still the hay molded and became dusty and unfit for the goats. Later we hauled it to an unused field where it remained for several years in mounds that slowly disintegrated into the soil.

The second summer the town manager called to let us know that the hayfields at the airport were open for bidding. Whoever won the bid had to have the hay cut and baled a week before the annual fly-in fish boil in mid-July. Sponsored by the local Lion's Club, this particular fish boil is an event to which pilots of small planes bring their families for the day or the weekend. The fish boil itself, of Scandinavian origin, is a meal of potatoes, onions, and whitefish or salmon, all cooked together in a cauldron over an outdoor fire. All are welcome to the meal, which takes place under tents at the airport. The airport itself is a modest affair consisting of a small building, a grassy runway, and usually a few small planes parked in a row. The hayfields that are part of the airport property are seldom used except during the fly-in fish boil, when tents are put up for the meal and the fields are made ready to accommodate all the visiting aircraft.

With our goat population then at more than fifty, Peter and I calculated that the airport hay would give us plenty, even if we were able to get in only one cutting of his own small field. I talked with John and Martin to get an estimate of how many bales we could expect from the field and how much it would cost us to have them brought in. Then we set the bid high, assuming that at least one farmer would bid in order to market the hay off the island. The town clerk called after the deadline for the sealed bids. We were awarded the work, she said; no one else had bid on it.

Then came the time to wait. June ended. Martin was to do both the cutting and baling because the quantity was too great for John's slower equipment. But when the weather was right, Martin had other deadlines. The town clerk began calling me almost daily, reminding me that the hay had to be off the field. The deadline came and went, and still Martin was busy. Finally he cut it. That year, because of the dampness, it was advisable to rake it, so John set me up with his old tractor pulling a rake. I drove through the windrows raking the hay all one hot summer afternoon. When it came time to bale, Martin's son and I worked the wagon, baling for several nights, the last one in a downpour the day before the start of the fly-in weekend.

We didn't have room to store such a large quantity of hay at the farm, so Russ, whose family built and lived on the farm for decades before Peter bought it, agreed to let us store it in the hayloft of his barn about a mile and a half away. I worked with John several afternoons getting the hay in, loading the conveyor while John stacked hay in the loft. Later that year, whenever I was out of hay John and I would move a couple of hundred bales to Peter's farm.

The next year we had Harvey cut hay. He's a commercial fisherman and a talented mechanic and was as busy, it turned out, as Martin; consequently our small hayfield was left until August. We bought other hay consisting mainly of grasses Harvey had cut, and stored it in the loft of Peter's large barn, but it was never very nutritious. Now I buy most of the goats' hay off the island from a farmer who buys good second- and third-crop alfalfa and hauls it over. It's expensive, and the hauling and ferry fees add to the cost, but it's a sure thing. The only nuisance is trying to determine a day when the ferryboat won't pitch too much for such a top-heavy load.

November

One November morning at three-fifteen I awakened thinking I'd heard something, or perhaps dreamed it. Then it came again, a faint catlike sound from the direction of the goat barn. I pulled on jeans, stuffing the tail of my flannel nightshirt into the pants, slipped on a sweater, and hurried outside. Too awed by the magic of the silent, shining night, I could not tell if it was cold. But the air was still and the clear sky had the quality of dark-blue glass. A nearly full moon shone in the west, its brightness masking the competing light of millions of stars. Thick frost coated the grass, causing it to crunch slightly under my step. So thick it looked like snow, it blanketed the barn roof and the top of the van and turned the spruce trees on the lawn into glistening Christmas-card firs. A month or even two months late, it was the first hard frost of the season.

I walked around to the barnyard outside the kids' pen and swept the area with the beam of my flashlight. Almost immediately I saw the problem: a large kid was hanging from the top of the wooden gate separating the kids' area from the does'. A hoof had caught in a slight gap between two vertical boards. I extricated him without much difficulty and carried him into a part of the barn where I could confine him for the rest of the night. The leg did not appear broken but nonetheless he refused to stand on it. Just a few days earlier I had noticed the very same kid, one of Ariadne's twin boys, chortling like a billy goat, his face grimy as if he had sprayed himself with urine as billies do during rut. I questioned then whether his castration a few months before was successful. Earlier I had felt what was left of his scrotum; it was hard and leathery as it should be after banding in September, and soon it would fall off. Possibly he had an excess of testosterone and the residual hormones prompted him to attempt scaling the gate to consort with a receptive doe. More likely, his testicles were not in the sac of his scrotum when the hard elastic band was slipped around it to cut off circulation, and he remained, technically, a billy with all the urges of mature billies when confronted by comely females just a wooden partition away. I left him and walked outside again to look at the sky and the newly frosted world.

It's a fall that people here will talk about for years. Jeanne at the bank remarked that even the old-timers maintain they can't remember such a mild autumn. By the midpoint of November there has not been a hard frost or even very much of any frost. Yellow, pink, and red snapdragons persist in the garden, and buds on the climbing roses still swell and open into small but recognizable flowers. The last rose of summer has become the last rose of November. Normally in late autumn my gladiola bulbs are drying, spread out on newspapers on the floor of the spare bedrooms. But by mid-month this year I still wait for the stalks to turn from green to yellow, signaling that nutrients have finished returning to the bulb and that they can safely be dug from the ground. This year the lavender at the back door faded two months late; the red and white petunias in the window boxes of the black barn linger, but just barely, as do the impatiens, finally mounding high around the house in their late season fullness. It's remarkable that they are not already covered with snow.

The second year with the goats the temperature also took us unaware in early November, but that time it slipped down in the night to fifteen degrees before anyone had time to adjust or prepare. Peter came to Wisconsin with his friend Tom, and I met them on a Friday morning at the ferry dock on the peninsula side. We drove off for a day of browsing in antiques shops in the county. Early snow had blanketed the picturesque villages along the bays, putting us in a Christmas mood, and at noon in a pretty restaurant we sat in front of a fireplace cheerful with gas logs, and sipped a pre-lunch Scotch and felt festive.

When I step outside on an early morning as in this mild November and marvel at the glass-clear sky still shining with moonlight and silence, I know these moments both fuel my time here and give it meaning. I wonder how I will manage when my life changes, as it must. The stillness sustains me; the animals ground me and restore my perspective.

Around eight in the morning of the day I extricated the male kid, I delivered a bag of apples from our tree to Toni. On the lawn the frost had melted away, although I had to work at scraping a thick layer from the van's windshield. On the way to Toni's the field where Martin pastures his cattle looked untouched by the night's brush with winter. An unaccustomed November sun shone nearly all month, and the delicately tinted sky, unlike the deep azure of the West, turned Lake Michigan blue-green, sparked with brilliant flashes where waves teased by the breeze met the sun.

When I lived in New Mexico I thought it odd that relatives in the Midwest always inquired first off about the weather and then told me in some detail about the temperatures and rainfall or snowfall or lack of it they were experiencing. In New Mexico the climate is generally pleasant within the parameters of the season, and if not it will surely change, probably within a few hours. Weather in the midsection of the country is more extreme and pervasive; it affects one's mood and circumscribes one's activities. It's like a perverse entity that enjoys flexing its muscles every so often, making us bend to its will.

So this November with its generous sun and temperatures seems a rare blessing. Cold grayness and ravaging winds are inevitable, but this year they are delayed. By eight-thirty most of the frost had melted. Glen came over to work on the roof of the new hay barn; Kim and Rick joined him and soon he was singing "Oh What a Beautiful Morning." Glen's dog, Tobey, and Kim's basset hound, Hildegard, romped on the lawn with Lily. Mary came to help me clean the goat barn. She'd take away a few truckloads of manure for the bulbs she was coaxing into their rocky property in the woods. We all mentioned the weather as if the act of giving it appreciation for its gift to us that autumn would appease it and help ward off winter's fury, perhaps even subduing it into further somnolence.

"Hope the wind doesn't pick up while we're working on the roof," Glen commented.

"Yes, but I wouldn't want to complain about the weather this fall," Rick replied.

No, indeed. No point angering the weather gods.

Turandot

A MILD AUTUMN blunting November's knife-sharp winds no less than confirmed a new ease in my life with the animals. It seemed that the previous years' trials had coalesced into a reliable fund I could draw from. I was no longer raw and nervous; my senses were fine-tuned to the goats; I felt equal to handling most emergencies in the areas of illness and birthing. And that year for the first time we had adequate fencing to keep the billies from the girls until I chose to breed them. Finally, too, my calculations of when to plan for kids were based on my own experience, not what I'd heard or read and merely hoped would work for me. I felt I could ask for little else.

That year I had decided to breed selectively to improve the herd and keep the number of animals in check. I identified the does I wanted as mothers and began feeding them extra rations of grain in keeping with a practice that's popular among goat ranchers. It's said that taking does off grain in the summer, then restoring it gradually about three weeks before giving the females access to the billy—or he to them—causes them to drop two eggs when they ovulate and thus increases the likelihood of twins. The special feeding, I have found, is not difficult to undertake. Instead of struggling among milling, pushing animals to separate the chosen few from the mob, I hardly intervened after the first couple of days but rather did little more than stand serenely aside. Goats that are singled out are quick to recognize they're among the elite. They crowd around the door to their pen at feeding time, then file through into the hay storage area where I give them their pails of oats and corn while their less exalted sisters hang back

The discriminatory feeding gave me increased opportunity to observe my favorites. It was then I noticed Turandot looked off. She was slow to crowd up to the fresh supply of hay in the manger, her movements almost languid for a goat. Always polite and not overly pushy when other goats rush in aggressively, her natural peacefulness now seemed exaggerated. She was lacking in animation markedly enough to cause concern. I took her temperature and was relieved but somewhat puzzled to find it normal. There was no sign of diarrhea, no hint of respiratory problems. I began taking her aside to feed her a small amount of grain earlier than usual. She ate it without enthusiasm. Or was I imagining it?

Preparations for the weekend relegated my concern over Turandot to the remote crannies of my mind. Kate and Ross and their children were scheduled to come to the island for their last veterinary trip of the year. Ross has a mobile veterinary practice and makes a journey to the island every two months between May and November to treat Islanders' dogs and cats. We became good friends early in my time in Wisconsin and soon Kate and Ross began bringing their young daughters, Julia and Olivia, and staying at the farmhouse with me. We have an easy routine. I watch the children while they see clients. Later Kate and I cook dinner together; Ross takes charge of pancakes and bacon for breakfast. We have noisy, informal meals, which are often interrupted by someone who forgot to make an appointment for a sick dog or cat or by other clients who simply stop by for a chat. The time for our own visiting is in the early morning, or later, if there's time, on a hike through the woods. At the end of the visit, I'm left with renewed serenity and forty-pound bags of dog food and cat food that they press on me.

Saturday evening I left dinner preparations in Kate's hands and slipped away from the kitchen to feed the goats and give Turandot extra grain. She ate, so I told myself firmly I would not worry. Besides, I was even more concerned about the tiny doe kid who was not growing properly. Recent fecal samples for the herd revealed coccidia and worms despite the worming just three weeks earlier and the five-day course of coccidiosis treatment late that September. These tainted samples had already changed my plans: Instead of proceeding with routine hoof trimming and pizzle rot treatment, I'd launched into another round of individual daily oral doses of Corid for coccidiosis and had just wormed everyone again. On the fifth day of the coccidiosis treatment I would scrub the kids' barn with Clorox and hose it down, since kids are by far the most vulnerable to the organism. As I planned my days, I tried to convince myself that Turandot's slightly troubling demeanor was simply a sign that she'd been brought down a bit by worms. Surely her spirit would renew soon.

I gave Turandot her regal name when she was our sole registered doe. She came to us among the group of twelve does that Susie Waterman had picked out several years earlier and had delivered to the island along with a second registered billy. I have little other experience with registered animals aside from the two billies, and cannot say with conviction that a registered animal will always have a finer, heavier fleece and better stance and body than a commercial-grade animal. I suspect that mediocre goats have made the registry simply because they had the right parentage. But if Turandot were my only yardstick, I would vouch for a registered animal. Her lush, curly coat, bright with luster, never failed to give her a weighted-down look right before its nine pounds were shorn off. And she was an easy goat, accepting my attention with dignity. Her calm attitude suggested responsibility, unlike the flighty Tosca, for example, and this was borne out in the attentive mothering of her kids. I never worried about her at kidding time; I knew we'd always be able to cooperate in launching her newborn and afterwards I could enjoy the kid while she did the work.

Postponing Sunday morning breakfast, Kate, Ross and the girls followed me to the barn to feed the goats. While Ross and I deliberated in the kids' barn, inspecting the tiny doe kid who seemed to lose ground daily, Kate, Julia and Olivia fed the others their breakfast hay.

"Look at that goat! Why is her tongue out like that?" one of the girls called to us from the does' barn.

Then I saw. A helpless-looking animal stood there, a hugely swollen tongue lolling out of her mouth. It was so enlarged that she could not possibly pull it in, much less eat or drink. I looked again. It was Turandot.

At that moment I was thankful beyond words that Ross was there. He felt around her jaw and throat. Both had swollen dramatically.

"Most likely the swelling is cutting off circulation to her tongue and causing edema," he suggested.

We took her temperature; it was a high one hundred five degrees, three above normal.

After feeding the goats, we returned to the house to research symptoms, paging through the various books I'd accumulated. I handed Ross a British veterinary manual on dairy goats while I took the two more general Angora goat books.

"There's sore mouth," I ticked off, "and something called blue tongue."

I found as I read on that sore mouth causes sores and scabs on the lips and gums. Adult goats are generally immune to it. It was not a likely possibility. The entry on blue tongue offered no better promise: an insect-borne virus that "rarely causes clinical disease"—but, one presumes, a blue tongue. Ross closed his book and leaned back.

"It's probably an abscess," he said. It was the obvious diagnosis, I realized. Like cats, goats are susceptible to abscesses from injuries but also, with goats, the abscess can be a sign of *caseous lymphadenitis*, a chronic bacterial infection. An abscess can be internal and when it is, there's generally no external clue. This, in fact, is often blamed for deaths when there is no apparent cause. I'd gained experience with this condition when I was called on to treat the abscess of a pygmy goat at the island's farm museum. The abscess had been lanced by one of the farm-animal veterinarians in Sturgeon Bay; my job was to go over to the museum every day for ten days to open the wound and flush it with hydrogen peroxide until it healed from the inside. It did indeed heal, gradually, and from the inside out.

After breakfast Ross and I returned to the barn to examine Turandot again.

"The abscess is still hard," Ross noted. "But it's beginning to soften. Probably by tonight or tomorrow morning you can lance it and let it drain."

"Lance it?" I looked at Ross. Can I do it? I wondered. I'd do it, I knew, but could I lance it without injuring her? What was the chance that I'd make her worse? I knew I wouldn't be squeamish about blood or draining the abscess, but what if I pierced a vein? What if she didn't stop bleeding? How would I even know where to cut in the massively swollen area that extended from the base of her jaw down her throat a good seven or eight inches?

Ross rummaged in his veterinarian's black bag for his electric clippers, then removed the hair from the area of the abscess. Again he searched his bag, found what he was looking for and handed them to me: a pair of forceps and a package of surgical blades.

"Wait until it's soft," he cautioned. "It may burst by itself," he added.

I studied Turandot. Clearly she could not eat or drink. I could squirt water into her mouth with a turkey baster but I couldn't help her with food. How long would she last in this condition? How long before the complicated stomach of a goat reacted to the absence of food, the lining turning acidic and sloughing off, as I'd understood might happen?

"I'll mix a little heper and echinacea," Ross remarked, explaining that heper is a homeopathic form of sulfa. "It will help process the abscess."

"Should I give her penicillin or anything else for her temperature?" I asked.

Ross advised against it. A fever is the body's way of creating heat to further the progress of the abscess toward bursting and ridding itself of toxins, he explained.

Soon afterward my friends left. Watching them drive away

I felt unmoored and I understood that it was the task ahead that unsettled me, the veterinarian's job of taking scalpel to flesh.

Evening fell and the abscess remained unyielding to the touch. I called Gene to see if he could come by to help me hold Turandot during the lancing procedure. I'd have to let him know when it was ready. In the meantime I squirted water into her mouth and hoped she would last until morning.

The next day it still felt hard except for the slightest softening at the top of the swelling. Ross again advised waiting. The day was busy. A truckload of hay bales, nearly seven tons, was scheduled for the ten o'clock ferry. I telephoned around the island to line up help to unload it so that the farmer delivering it could get back by the next boat. Also, Dick was to arrive in the late afternoon. I'd planned a dinner party so that he could meet Joan and Clay Blair, who live on the island and are authors of histories of the Second World War and the Korean War. Because of my weekend houseguests I'd not yet shopped for groceries for the evening's dinner, which on the island is very bad planning indeed. On Mondays the market's shelves are all but bare; meat, vegetables, and most supplies arrive weekly, on Tuesdays. At the grocery there was not so much as a single drumstick of chicken and only a bit of tired broccoli, a few heads of iceberg lettuce, and not much else. So taking stock of what did remain, I replanned my menu, then hurried back to the farm to administer the third day of the five-day coccidiosis treatment.

Normally, working with goats has a soothing effect on me, and I relaxed into the rhythm of catching goats and steadying them. I felt pleasure at the contact with their warm bodies and soft hair, at their quizzical looks and sprightly antics. Suddenly, I felt a kind of snap in my hand and something gave way in my grasp. The horn of the young goat I held had broken, not off completely but enough to be loose at the base. I looked down and saw she'd begun bleeding from the mouth and nose. I ran back to the house, wondering once again why we'd never installed a telephone in the barn, and called the Peninsula Veterinary Service. Dr. Haas, one of the vets there, reassured me it was no cause for alarm. There was nothing I could do about it anyway, he informed me, because we lacked equipment for dehorning the goat and cauterizing the wound with heat. In time the horn would probably harden in place, he added. I mentioned Turandot and my inability so far to find a soft spot in the abscess; he suggested I use a needle to locate the best area to drain. And when I did cut to drain it, he cautioned, "stick to the midline" of the swelling.

It was a relief to immerse myself in dinner preparations. It was a respite, too, to entertain friends, letting their conversation and their world of journalism, publishing, and history distract me. When I checked the barn at eleven-thirty that night I found the tiny kid snug in the kidding pen where I'd removed her to keep her warmer. Turandot looked much the same as before and it seemed wise to postpone any treatment until morning, when I'd be fresh.

The next day I still didn't feel ready to lance the abscess, nor did the abscess seem any more ripe for the cut than it did earlier. I called Ross, ostensibly for that elusive bit of information that would make it easier; mainly I wanted support. Ross, too, suggested first probing the still-hard abscess with a large needle to determine the best place to drain it. I gathered cotton balls, alcohol, hydrogen peroxide, and iodine; I ran warm water into a pan and added a blue disinfectant called Nolvalsan. I took a large sewing needle from my supplies, the individually wrapped scalpel blades and the forceps, and a couple of pairs of surgical gloves. With these, I headed toward the barn.

It was easy to spot Turandot: she was the only goat lying down on the dry November stubble. The goats began moving toward the barn; Turandot, weaker than the others, lagged behind. I guided her into the hay storage area, where I'd set up my supplies on a wooden chair. Dick was close behind me.

"If you feel faint or think you're going to get sick, let me know," he offered as I showed him how to hold Turandot by her horns.

"That's not the problem, Dick. I'm just afraid of hurting her. I'm afraid I might somehow do real damage to her."

I wiped the needle with alcohol and inserted it at the base of the abscess. It seemed to go up through dense tissue to what felt like an opening of some sort. I withdrew it and pressed the surrounding flesh to encourage the draining I'd hoped for. Nothing happened. I cleaned the needle again and tried another spot. After five or six tries there was still no result. I felt easier about the procedure, though—more like a veteran with the goat apparently unharmed. I checked Dick's watch; it was almost time to drive him to the ferry dock. I'd have to try later, perhaps with the blade. I'd call Gene first.

Gene was skeptical of the efficacy of holding a scalpel blade with forceps so he brought a straight razor and a hunting knife to use if the other didn't work. I tried the scalpel blade first, but it merely wobbled in its holder when I tried to cut. Using Gene's hunting knife I made a small incision, again at the base of the abscess where it would theoretically drain more readily. I kept to the midline of the throat. Not knowing the map of veins and arteries that presumably crisscross the flesh under the surface of skin and hair, I feared slicing through major vessels. I pictured blood gushing.

Motionless in Gene's grasp, Turandot showed no sign of pain. Is the tissue dead? I wondered. I could feel the blade slice through something tough, then through tissue that yielded easily to a point at which it was unaccountably difficult to make the knife cut farther into the flesh. Was it merely my inexperience that impeded it? Suddenly the tip of the blade moved easily. I withdrew it, surprised by the meager amount of blood around the wound. I pressed to drain the abscess, but nothing happened. I inserted the knife a second time to make the opening larger. Still nothing happened.

With Gene holding Turandot, I returned to the house to call the vet.

"Make the incision large enough so that you can get your finger into it and into the cavity," he directed. "Then you can work it around to start the draining."

Getting myself geared up for that second assault wasn't easy, and I was grateful for Gene's calm. As I poked my finger into the wound I could see the tissue in my mind's eye: pink and smooth and folded over in places. An inch and a half in, I'd gone past the place where I thought I'd felt the cavity before. Now there was no opening, just dense tissue, so I decided to try cutting at the slightly softer spot I'd felt near the top of the abscess. Turandot winced the second time I cut into her. Nevertheless, I felt more confident, less tentative; mainly, I wanted it to be done with. Still I marveled at how a blade in my own hand could slice through living flesh of an alive, fully awake being. There was no need to make the incision large enough for my finger to poke into it; with a little pressure, thick, bright red material oozed out. The abscess was surely draining, the contents colored by blood. I injected hydrogen peroxide into the incision and pressed again; the abscess drained again, the material frothy. I repeated this several times.

Turandot showed improvement immediately; the swelling in her throat did not subside dramatically, but apparently it was enough to relieve some of the pressure. Gene still held her horns but she managed to stretch enough to reach a nearby bale of hay and nibble. I love the basic, uncomplicated quality of animal care: Remove or alleviate the problem somewhat and you're rewarded by an encouraging thrust toward normalcy. Turandot's tongue still protruded from the side of her mouth, but she looked better. It would be only a day or two before the edema subsided, I thought, feeling jubilant. I'd keep the incision open and drain the abscess two or three times a day. Gene continued to hold Turandot while I set up a kidding pen for her.

That evening Turandot ate some grain, although much of what she tried to take in got wet and slimy and fell to the ground. But nonetheless, she was eating and I was relieved. The next day she continued to nibble at hay and grain and water administered by turkey baster. By evening her tongue was noticeably diminished in size so I poked it into her mouth where it stayed. A good sign, I thought. At the same time, though, I saw she was not eating, or at least she showed less interest in food than she had the previous evening. Several more times I opened the incision and injected hydrogen peroxide. But I failed to drain the abscess again, nor did it reduce further in size. Turandot politely let me poke and this, too, concerned me. I would have felt better if she'd shown some feistiness. But then, I told myself, it's not her nature.

My relief was short-lived; my concerns proved apt. By Thursday she was worse. One side of her face was so swollen that her nose looked distorted; her temperature had plunged to a subnormal one hundred and five-tenths of a degree, alarmingly low. I called the farm-animal vet; I'd take the nine-fifteen boat and would be in Sturgeon Bay by eleven. As I lifted Turandot into the back of the van, I noticed she was much lighter than I expected. I shoved in a bale of hay; then, as an afterthought, I added the small kid who was not growing properly. I'd have her checked, too.

Dr. Dietzel was there when I arrived. He took Turandot's temperature first off; it had dropped to one hundred degrees. He felt the abscess.

"It's meaty, not like an abscess but more like a tumor."

He listened to her heart. There was a pronounced heart murmur and edema in the chest area.

"We can take a sample of the tumor and biopsy it," he suggested. "But chances are she won't survive. With the tumor where it is, she can't eat. She'll starve to death soon."

The heart murmur itself was a bad sign, he explained. It indicated pressure from a tumor or an internal abscess in that area, an additional growth that was not necessarily related to the one along the neck. It would be best, I decided, to euthanize her. I felt like two people, one making the rational decision to put her down, the other objecting that my favorite goat apparently was fine only days ago, that this simply could not be happening. With Turandot still in the back of the van, the doctor clipped her fleece in order to find the jugular vein. He administered a lethal injection directly into the vein. She slipped away as gently and discreetly as she lived.

After Dr. Dietzel examined the kid and found nothing apparently wrong, we talked about Turandot. I pointed out the luster and unusual curl in the coat of an animal nearly five years old. I'd had other animals—dogs and cats and other goats—euthanized by vets who are unfailingly respectful and grave. They must see death all the time, yet it's never casual to them.

I drove home with the little kid nestled against the dead doe in the back of the van. I buried gentle Turandot with the others, by the stone fence just beyond the west side of the pasture. Although it was almost mid-November there were enough snapdragons and lavender still blooming to put in her grave.

Little-Little

LITTLE-LITTLE is dying. She may be lying stiff and lifeless in her pen as I write this. If not—if she's survived the night—I will figure out a way to bring her into the house to spend her last hours in my company, which seems to soothe her. This won't be easy; generally, the dogs are only mildly curious about the goats, but Molly once began salivating at a newborn kid when I had to bring it into the house. I wouldn't trust her with an older kid, either.

I don't know what is wrong with Little-Little; I know only that she will not live long. I know this because her temperature has dropped below normal, remaining at one hundred degrees and sometimes dipping below when it should be one hundred and two or more. And there are other signs. A week or so ago, whenever I came into the kids' area of the barn I would find her stretched out on her side, either motionless, her eyes closed, or crying out in distress. She apparently had been knocked over and could not get up. After this happened several times, the only solution was to put her in a kidding pen under a heat lamp where she would be safe from other goats and more comfortable with the added warmth. But goats are herd animals, sociable animals, and don't do well in isolation.

For two nights I put Nippy, a friendly little kid, in with her. At first it worked. Nippy took to the change of venue cheerfully, and with great relish munched the green alfalfa that was all his for the taking. Perhaps heartened by his vigor, Little-Little ate with more purpose than I had observed recently. I felt encouraged. But the second morning I came out to find her lying helplessly in the hay while Nippy pranced on top of her in his eagerness to greet me.

I tried putting her mother in with her, but without success. It was apparent the young adult had no remaining interest in her kid and was unlikely to stay contained in the pen long enough to rekindle any latent recognition. When I led the doe back to her area, Little-Little followed us, bleating forlornly. I was surprised she still had sufficient strength to do this; surely it was the strength of the bond that remained for her but not for her mother. The image of the forsaken baby still tugs at my emotions.

The rest of the day Little-Little lay rag-doll limp in such odd positions that I was certain she would not live until night. I considered digging a grave. Wind blew from the northeast with gale force; snow was expected. The weather would only grow worse and digging a grave now seemed prudent, but I could not. Call it superstition or the last vestige of hope, but something usually prevents me from making a grave before an animal has actually died.

I had to spend much time in the barn while the wind raged outside; there were odd jobs to do, including separating the goats so that female kids would not be impregnated inadvertently. Once I removed our six doe kids safely into the area with their mothers and other older does, I had to make arrangements so that the little females would have access to food without being crowded out of the way by the larger animals. All this kept me in the barn, back and forth between the different areas. Whenever Little-Little heard me approach she would bleat until I came in. Sometimes, when I was very busy, I simply called out to her, which seemed to satisfy her. Other times I climbed into her pen and stayed with her for a while. Then she acted alert and nibbled at hay, although I'm not certain she actually ate it. She would have been happier, I'm sure, if I had remained with her, crouched in her pen, talking to her and stroking her. I began to ponder how I could bring her into the house yet protect her from the dogs.

Ideally, Little-Little would not have been born, at least not this year. She was the product of my wish to keep one strong boy kid as a billy when I had no room to separate him from the rest of the kids. I reasoned that if the doe kids were not ready to breed, they would not conceive. But two did conceive, and motherhood coming too soon can stunt a young female's growth because the fetus saps the calcium the mother needs for her own development. And there was a question of whether her milk would be sufficiently rich in antibodies and nutrients to protect her newborn adequately.

On the afternoon of April 14, Carmen's kid daughter, Pilar, began labor. The little male baby was small, and just a couple of tugs from me helped it slither onto the hay. Two and a half hours later, on the same spot, Sophie went into labor. Again I was thankful that the birth was uneventful, that the tiny female baby was small enough to emerge easily from her young mother. And Sophie, although still a child-goat herself, immediately demonstrated all the appropriate maternal instincts that some older does need time to develop. She licked her baby dry while cooing, nudged it toward her udder and in general was attentive. I kept both pairs of kid does and their infants in kidding pens longer than the usual five days because the babies were so small. The male had a precarious beginning. A rattle in his throat signaled fluid in his lungs; the vets call it mechanical pneumonia. At one point he would not suckle and I could not easily get a stomach feeding tube into his esophagus. I held him and inserted it, then listened and heard breathing through the tube, which had slipped into the trachea. I tried inserting it repeatedly and wondered what damage I might be doing to the delicate lining of these passages. Over the telephone the vet explained that his difficulty with breathing caused him to hold the entrance to the trachea open, making it a ready avenue for the feeding tube. Finally I was successful; two feedings of warm milk gave him strength to nurse on his own again. Some penicillin and steroid also helped him along; he recovered and gradually caught up with the other goats in size. Sophie's baby was apparently healthy from the first and I did not worry about her. Although her grandmother, Jane Austen, an aggressive, cantankerous goat, was large, Little-Little took after her mother in size and sweetness of disposition. At least I assumed her petite size was a normal, inherited trait, forgetting that the mother's growth might have been curtailed by the early pregnancy.

During the summer Little-Little seemed to be simply a small and very friendly kid, one who could be picked up and who liked to be held. She nibbled my hair or collar and looked at me with a solemn gaze through her long eyelashes. She was too small to reach the grain trough easily when she and the other kids were separated from their mothers to live together in their own barn. But she simply hopped onto the feeder and ate while standing in the grain. This is something to avoid because of contamination of the food, but in Little-Little's case it was a workable solution. I did not worry about her, although one visitor remarked that when a very small lamb of hers was ultimately butchered, it was discovered that she had only one kidney. This, they suspected, accounted for her failure to grow.

By October Little-Little was markedly smaller than the rest of the kids. She looked more like a goat of two months than one of six months. At that time, a fecal sample indicated worms and coccidia among the kids and does despite recent treatment for both. I wormed everyone again and administered five days' worth of individual oral medicine, then scrubbed the barns with disinfectant as well as I could. Was coccidiosis the reason for Little-Little's size? I wondered. It is caused by a single-celled organism, with the goat variety specific to goats and found in all herds. The organisms, or coccidia, flourish during times of stress and in kids can damage the cilia that line the intestines, preventing adequate absorption of nutrients and ultimately stunting growth or causing death. Usually the symptoms are diarrhea and loss of appetite, neither of which were problems for Little-Little.

Ross had examined her when he and his family visited for his last vet trip of the season. Probably not coccidiosis, he'd ventured, but something is not right. Shortly afterward, when I took Turandot to the vet in Sturgeon Bay, I took Little-Little along. Her heart and lungs sounded fine, according to Dr. Dietzel. Perhaps vitamin shots would help.

Realization comes in degrees. Slowly I began to accept the idea that this sweet animal would not survive. I still hoped for a miracle but at the same time I rationalized that early death was perhaps better. A herd animal's quality of life is poor if it cannot keep up. Even if I could pen her with smaller contemporaries, she would remain the one subject to abuse and crowding from food. But as a herd animal she would not have a satisfactory life away from her own kind.

The next morning I again wonder whether Little-Little survived the night. After feeding the goats, I take Little-Little's temperature. Now it's even lower: ninety-eight and six-tenths. I take it a second time to be certain. I know I will work outside the house most of the day, and because I've not resolved the problem of how to keep her safely in the house, I put the next smallest doe kid in with her for company. As I lift the newcomer into the pen, I'm momentarily surprised at her weight, so much more than Little-Little's. She seems to welcome her new companion, who tries to climb out at first but then settles down.

I don't know what will happen to Little-Little. Perhaps for her sake it would be best to let her die. If I were to leave her with the other kids, she surely would. She would be knocked over, would not get up, and with no assistance from me, would eventually succumb. Even now I encourage her to eat and I keep her warm, both of which prolong the days of a little creature who is not strong enough to survive. Am I being cruel in the long run by taking measures to make her more comfortable for the moment? Would it be kinder to take that hard step, to deny her food and warmth? The fact that I do not separates me from real farmers and ranchers, I tell myself. Yet when I see burly, rough-spoken goat shearers on their knees in a kidding pen, holding a newborn kid that won't suckle up to a teat, speaking to the kid softly, affectionately, I am sure that my impulse is not foreign to more experienced hands at this business. What it comes down to is a response to life as if there's something in us humans that is programmed to nurture it. Dick, who likes to point out to me that there are too many goats in the world and that by raising goats I make no real contribution, understood this response when he helped me treat Turandot. So now, even as Little-Little is dying, I do what I can to make her comfortable and in doing so, make myself more comfortable emotionally.

Goats are tenacious. They'll work at a fence until they tear it apart; they'll repeatedly butt something that moves ever so slightly just to see if they can get it to move a bit more. And even as they are dying, they cling to life.

Every night I sleep uneasily because of the heating lamp dangling over Little-Little's pen. When a barn burns down, the fire is often caused by a heating lamp. Little-Little's temperature continues to drop. At ninety-six degrees she still eats sporadically and seems alert, although weak.

Tonight I bring her into the house and wrap her in towels warmed in the clothes dryer. I sit by the woodstove holding her, a bundle of light blue terry cloth in my lap, a little goat nose peeking out, resting on my arm. Lily tries to nudge her out of curiosity. Molly, her eyes huge opaque pools, looks stricken at the creature taking my attention. Am I forsaking her for a small goat? I give Little-Little a few ounces of warm milk from a bottle. Long past the point of taking a nipple, she merely swallows the milk without making an effort to take it on her own. I don't know whether at her age she has lost the ability to nurse properly or whether she simply doesn't want it. At least I'm warming her on the inside as well as the outside. After a while she relaxes in my arms and sleeps. I place the small bundle on a chair in the bathroom, safely shut away from the dogs, while I prepare hot cereal and a large cat-carrying case, stuffing it with hay for bedding. The cereal is a partial success: She swallows gobs I stick in her mouth with my finger. She clambers out of the carrying case, though, and it seems better to take her back to her kidding pen.

The next morning her temperature is up to one hundred

and one degrees but she won't swallow cereal or even nibble at grain. By the next day, she's slipped into a comalike state, lying on her side, eyes closed some of the time. She bleats when I come into the room, and it seems each bleat calls to me not to leave her. I respond, but all I can do is stay for a while.

For two more days she lies hardly moving, her head turned back so that her nose rests on her side. In my experience, which is still limited, this is the position goats are usually found in when they die. Still she cries when I come near. On Saturday I dig a grave; the weather has turned mild again, probably for the last time before the true cold sets in, and the ground is still soft. I expect to find her dead the next day, but she still breathes. I call Ross, asking if I should use the horse tranquilizer I have on hand to put her down. Ross advises against it. It's easier for her to make the transition naturally, he says. I don't know whether he's speaking of a physical or metaphysical transition. Perhaps both. I've witnessed at least one difficult job of euthanizing a goat, and do not want to risk additional hardship for her.

She died during the night. I was not there. For several nights before she died I thought of keeping a vigil in the barn, certain she would go at any moment. But she hung on, and seemed more peaceful when I was not near because only when I approached did she cry out. I hope she slept when I was not there, and I hope she died in her sleep.

Late Breeding

It's mid-January and I'm elated. My favorite of all our does is flirting with Jose. As winter wore on, I began to think I'd missed my chance this year to breed Ariadne. I've opted to pair the billies and does selectively and no longer let our current bucks, José and Radames, run with the females. So I must notice when a female is ready. In November Ariadne stood by the fence, shaking her tail demurely at the billy. "I'm interested if you are," she seemed to say. Of course Jose was interested and let her know, chortling, curling his upper lip, flicking his tongue and pawing at the fence. But I felt it was too early. The shearers were scheduled to come the first weekend of April; I wanted to delay birthing until mid-month. Ariadne's next period of fertility theoretically would be in roughly about twenty days. After two and a half weeks I began watching her closely. She's a beautiful goat, in disposition as well as in looks. She's gentle and easy and has a consistently heavy fleece that's retained good curl despite her six years. At one spring shearing her fleece weighed in at twelve pounds, an astonishing weight, particularly for the March-April clip when fleeces are lighter. Besides having good hair, she's an attentive, solicitous mother. I never worry about her kids either during birthing or afterwards. And I simply like her and the way she stands quietly gazing at me with solemn brown eyes while I stroke her cheeks.

Once in December I thought she was ready. I put her into the pen with the billy, but she seemed distressed and disoriented and clearly not enthusiastic about a romantic tryst. To a female who is not cycling, a large male must seem insufferably boorish. In her effort to escape José, she was cornered by him and by her son Radames. The two began sparring while Ariadne huddled between them. I quickly led her back to the girls' quarters.

The other goats I wanted to breed each became receptive in time: Carmen, Antonia, and Colette. Just last week I noticed Desdemona out in the field with an amorous wether in persistent if futile pursuit. Turandot's death in November left her daughter, Desdemona, the only doe to carry her good genes, and I'm eager to have her offspring. I couldn't risk putting her in with the wethers and billies because José is most likely her father and in his dominant position would certainly assume the *droît de seigneur*.

I've read that animals in the wild avoid inbreeding unless the population is circumscribed and limited. Perhaps centuries of domestication have relieved our livestock of the burden of making these finer distinctions themselves. In any event, goats don't show much discrimination. I've never noticed any sign that a billy goat has any parental awareness at all. To a lusty stud, any female is foremost a female. But I've noticed, too, that when a billy goat is thwarted by separation from his harem, deluded and perhaps filled with hope, he will smack his lips over the back of a comely wether.

I pulled Radames from the boys' dorm and tugged at him until he finally trotted along into the boy kids' area. Then I caught Desdemona by the horns and led her in, too. Youthful, rambunctious lovemaking! In and out of the barn Rademes pursued Desdemona. Into the pasture despite snow, then out again. Coy retreat, then lingering encounter and retreat once more. I left them for twenty-four hours with the wethered kids who scattered whenever Radames charged them, as if they could be a threat. The next day I removed Desdemona and Radames to their respective pens.

In the meantime, I kept waiting to see Ariadne sidle up to the fence and flag her little tail at the billy. Mentally I asked her if I wouldn't get another chance for a daughter to carry on her lineage. Every morning my gaze swept the pasture to see who was standing apart looking over at the boys' area rather than milling around me for fresh hay. Sometimes it was one of the doe kids, far too small to breed; other times it would be females with inferior fleece.

January is late for breeding. The books on raising goats in northern climates say the breeding period is from August through December. But today, January 13, Ariadne stands at the pasture fence, shaking her tail at José. She doesn't wag it vigorously but there is a definite show of interest. It's worth a try. I take her by the horns and lead her out of the girls' area, through the hay storage section stacked high with bales, and into the boys' pen where José, Radames, and the adult wethers reside. Having momentarily lost sight of Ariadne, José stands in the pasture doorway, appearing puzzled until he spots her. He waves his massively horned head from side to side, then points his nose in Ariadne's direction and charges toward us. I give Ariadne a quick shove into the pen and retreat, slamming the gate behind me. Rolling his eyes, snorting and stamping, flapping his tongue and snuffling, José reaches Ariadne. He's a mess with his face and forelegs doused with urine, smelly and blackened by the grime that adheres after a thorough drenching. José sniffs Ariadne for a moment, then points his nose in the air and curls his upper lip. Finding her scent to his satisfaction, he smacks his lips up and down her back. This time Ariadne is pleased and tries to escape only half-heartedly, making him work ever so slightly for her favors. He begins mounting her, quick thrusts, then he's off and pursuing her again. The two dance into the pasture, a *pas de deux* of goat lust, then back to the barn.

Later I feed them. Side by side, José and Ariadne look like a comfortable married couple. Amid the melee at feeding time, Ariadne takes her place serenely next to Jose and munches corn and oats that others have to push and shove to get near. It seems as though José is consciously protecting her from the struggle, clearing a place for her so that she can eat her fill. I offer them some of the best quality alfalfa, which I generally reserve for the does and kids. Again next to Jose, Ariadne takes her share peacefully. She's oblivious to the furor in the barn, insulated by José's charisma. Occasionally Radames moves too close; José bellows and lunges. No need to worry about inbreeding here, for José refuses to leave Ariadne's side.

The next morning I come into the barn before dawn. Jose and Ariadne are settled in the hay, side by side, peaceful and perhaps tired after hours of romantic gamboling. But when I feed the goats, I notice José is no longer Ariadne's protector. She must now fend for herself. It's time to take her back to the girls' pen and hope for results in five months' time.

April

An April snowstorm swept down from western Canada, picked up additional moisture and cold as it hurtled across Lake Superior, and in the last hours blanketed the island. It will pass, I know, and the snow won't remain; after all, it's April. I'm hopeful in April, which seems like a season unto itself, not winter but clearly not yet spring. It's still bleak and cold and the relentless wind now makes me strain toward a mental landscape of green shoots pushing through moist earth, a place of warmth and a soft breeze. Earlier, in March, there was a glimpse of spring when the temperature climbed briefly to forty degrees. Shortly afterwards I found the bodies of hundreds of bees around the hives and as far away as twenty feet. They had flown out, deluded by the sun and foolishly looking for nectar, and were chilled before they could return.

I'm leaning toward spring, but I don't know where it is. I'm looking for it as if it's hidden and simply requires my finding its secret place or sign, as in a fairy tale, so that the icy spell can break and magically it can smile its warmth on the landscape.

On the island spring becomes a long, empty craving throughout days as raw as winter but grayer and bleaker than ever experienced in the crisp cold of January and February.

Goat Song

Bare trees, brown fields still dry and hard, gray sky, north wind: April on northern Lake Michigan.

Yesterday I deloused the goats using Expar, which makes an oily line down their backs and causes the short hairs of the recently shorn animals to separate so that pink skin shows. It causes a syringe to stick so I use a bulb baster, the kind for basting a turkey. I did the kids and does and afterwards dusted each animal's underside with diatomaceous earth. I left the boys for today. It takes extra energy to do them. I will pull each one into the hay storage area so I can delouse them without having to be wary of the billy, because he's possessive of his herd and likes to challenge me when I interfere. When he intrudes I rap him sharply on the horns with a plastic pipe to remind him to keep his place, but he generally comes back for more. The job of delousing the boys is strenuous enough without battling José. Each wether tends to dig his hooves in and refuses to budge. If I try luring one with grain, the others crowd in for a treat.

I gear up mentally to work with the billy and the wethers. It's not that I'm afraid of our billy goat, but I'm very much aware of his potential to do harm. Some women goat farmers refuse to keep billies because they're too difficult. I choose the word difficult over the word dangerous with intention; I don't want to risk letting fear lodge itself in my perception and hamper my work. I've been knocked down by the billy, butted when my back was turned. One's movements around goats become finely tuned, and I have a sense of precisely how long I can turn my back depending on where the billy goat is standing. Likewise, I know just how long I have before the does or wethers will crowd through a partly opened gate. The time the billy succeeded in butting me to the ground; I misjudged and thought I had moments enough to turn to dump hay into his feeder. He butted me in the back of the legs; I fell to my knees and immediately turned to find him coming at me again. I grabbed his nose with one hand and began slapping his face with the other. This deterred him just enough for me to get to my feet and out of his reach.

For this morning's delousing I'm using a syringe to assure myself that I'm not overdosing with insecticide. I pull the goats into the hay and grain storage area, two at a time. The boys are just curious enough to cluster around the door of their pen so that I can grab them by the horns easily. It's always the same ones first: Riccardo and Phil and little blue tag No. 30 and yellow tag No. 18, Elsa's son. Just as certainly I can predict which ones will hang back and ultimately require snagging with the crook.

It's a contest of strength and will, but I enjoy the physicality of the work, the sense of using my arms and legs well and having the strength to do this task. After an hour and a half I've nearly half completed it. I fleetingly wonder why I dread this operation so much that I put it off until the last, whereas I'm eager to work with the does and kids.

Just as I'm dusting a big three-year-old animal I look up and see the billy and a group of large wethers crowd through the gate and into the area where I'm working. The gate apparently failed to latch securely because of all the hay trailed underfoot. I grab my PVC pipe—my billy club—and jump toward José, seizing him by the horns. Before he has a chance to react, I pull him back into the pen and shut the gate. Then I begin grabbing wethers, shouting at them and shoving them, hauling them by the horns into the pen. Ordinarily the way to get a group of goats to cooperate is to lead them with a bucket of grain. But I know that with this crowd of twenty-five animals milling around the storage area, if I so much as lift the lid on a grain bin, at least a dozen 150-pound wethers will overwhelm me as they push into the feed. By this time the goats are ecstatic. They're inquisitive animals and intelligent enough to delight in a change from the normal routine. They hop onto bales of hay and climb the stacks; they poke into corners, knock over empty grain bins and lick out bits of soybean meal left in the plastic containers. I'm angry by this time, not furiously angry but just enough at the inconvenience to muster a useful amount of adrenaline. It's one thing to drag a goat to his pen, another to get him through the gate and back into his part of the barn when he's been having a good frolic. It takes a determined shove from behind while holding the door open only partially lest the animals on the other side surge through again.

I've almost succeeded in restoring order. I'm tired by this time but not so much that I can't finish. Then I look up and see that it's all repeating. The billy, the whites of his eyes showing wildly, waves his massively horned head from side to side. He charges across the area; I run to meet him, grab a horn, and drop my pipe. He charges me as I hang onto his horns and go partially down to reach for the pipe. He pushes me back against the one full feed bin. Still clinging to his horns, I reach for the pipe and miss. I slap him as he tries shoving me with his head, then I reach again for the pipe. I grasp it this time and hit him as hard as my half-crouched position will allow. It doesn't faze him. I straighten up, never letting go of his horns, and wrestle him toward the door, lurching with him as he repeatedly tries to charge me. Somehow I get him in.

I turn back to the others. To my horror they've discovered the bin full of feed. I'm horrified because I know goats will eat grain until they're so bloated they die, and five are now pressed into the feed bin headfirst. I race to the bin and begin trying to pull them off. They're all clambering around me. I pull one off and another takes its place. I'm shouting, yelling, pulling at horns, shoving sturdy, well-muscled animals that are as determined to eat grain as I am to prevent them eating it. Somehow, they seem to have grown; they all look big to me now. I manage to slip the lid partially onto the bin and simultaneously beat back enough goats so that I can push the lid down into place. A full grain bin contains nearly two hundred pounds of feed. This one is nearly full. It takes all my strength to move it. I drag the bin, fighting off goats, until I get far enough into the maternity area that I can close the door. I hope the two does in kidding pens with their babies won't try to investigate. Next I haul the remaining sack of feed, which weighs somewhere around seventy-five pounds, across the floor of the barn and out the door to the outside.

Panting from exertion, I go back to dragging goats into the pen. But this time the billy is determined to break out at the first opportunity. He stands at the gate butting any goat that I shove in, trying to butt me and get through the door. I'm running out of strength and he's too much of an obstacle. If I can get him into the pasture and somehow secure the door—which no longer latches properly—between his pen and the outdoors, he'll be outside where he can't harass me. Armed with the pipe and carrying baling twine to tie the door shut, I enter his pen. He's ready for me. He's angry and determined. I grab his horns and he charges, pushing me back a few feet. I hit him on the horns with the pipe to subdue him, remind him, but he's past noticing it. I hit him on the rump; it doesn't impress him. Now I'm against the wall, the billy pushing me with his head. Shouting and beating at him, I struggle with him toward the pasture door, but it's no use. He's determined to fight me, and even if I get him out, I'll still have to manage tying the door shut while he butts and shoves to get in. I work my way out of the pen, the billy trying to butt me the whole way.

Safely out of his enclosure, I gulp air; my throat is raw from the dust kicked up and from shouting. Wethers—twenty-five of them—are everywhere, and the billy is waiting for them at the gate. I'm without reserves; it's the first time I haven't had strength to finish a task. Either I must rest or get help. I slip out of the barn and call John's house. It's now noon and he'll be there, but the line is busy. I call another friend but get only the answering machine. John's line is still busy. Then I think of Lyn. He and Nancy have horses and cows, and he's likely to be home. I call and he says he'll be over in a few minutes. I go back to the barn and resume delousing, grabbing goats as I find them and running a line of Expar down their backs.

Lyn comes and calmly suggests a bucket of grain. Why hadn't I thought of it? I wonder. But by that time, exhausted and having moved the grain bins to ensure the animals would not have an opportunity to gorge themselves, I did not give the feed another thought. It works beautifully with José, who seems puzzled by Lyn. With the billy shut out in the pasture, Lyn and I easily shoo the wethers into the barn. I'll leave the billy outside for a few hours to settle down while I finish delousing the wethers; I don't have the energy to dust them with diatomaceous earth. I'll do that another time, and I'll delouse José tomorrow.

Yesterday I left my goat work earlier than usual in order to shower and change for a funeral. Dale, from an old island family, died suddenly Monday morning; a friend who was his neighbor told me when we met to walk that Monday afternoon. Pat, calling as part of the Lutheran women's prayer chain, advised me that the service would be Wednesday at one o'clock with a light meal afterward. I knew Dale only slightly, mainly through John's great regard for him and his work on the Norwegian-style *stavkirke*, or stave church, being built as a Lutheran chapel by John and members of the community. Whenever I met Dale at a stave church committee meeting or ran into him at the grocery store or lumberyard, he had a kind smile and a few words of inquiry to show his interest.

The sense of community is never so strong here as in time of grief or trouble. If someone falls seriously ill and may not be able to afford expensive medical care, islanders stage fundraising events—potluck meals and raffles—to muster the needed cash. Prayer chains, mainly among women, circulate for those who are sick; a roster of those ill or grieving is read each Sunday at church and it makes not a bit of difference which church the troubled attend or even whether they attend at all. Like many churches, Trinity Lutheran is crowded on Easter Sunday and Palm Sunday and a few of the other notable days of the religious calendar, but at funerals it overflows. Chairs are set up in the narthex, people crowd the stairway at the entrance or stand outside. Because I hardly knew Dale, I hung back, not wanting to take a place inside when others who knew him would have to stand. By twenty minutes before one, cars and trucks crammed into all the available parking places around the church and had nearly filled a little-used field to the east. Most people were already seated, others milled around outside, the men formal in jackets and ties, rarely seen on the island except at weddings and funerals, the women dressed for Sunday service.

When Colin died, the school gave all the children a half day off in order to attend his funeral. When an islander dies, almost the whole community takes time mid-day to attend the service and the lunch afterwards that is provided more or less spontaneously by the women of the church. It's a matter of making time and space to feel the communal loss. It's these times that I understand the value of what I will give up when I leave the island. I won't become a part of the community in the way that Dale and Colin and so many others have been. It's my loss, and I value what I'll miss.

Late Kidding

Some of the does' udders ballooned days ago and look like large inflated sacks suspended in front of their back legs. They stand awkwardly and waddle, their hind legs apart, or lie in the hay, grunting. Female goats appear philosophical to me these days. They seem to look inward, as if unconcerned about outward things except at feeding time. I imagine that they somehow know that their purpose at this time, the kids growing inside them, relegates all else to mere trifles, and this gives them their transcendent calm. A normally meek goat suddenly is unaccountably aggressive. Is this new twist of personality hormonal? I wonder. Is she feeling unaccustomed strength surge through her in anticipation of the offspring she'll protect? Perhaps her self-esteem has risen with her new mission. My favorites pull themselves to their feet when I come in and amble over to greet me.

So far during this fifth kidding season four goats have been born, three days in a row. Three were too early according to my calendar, and the twins were so small and weak that I wondered if they were premature. The brother and sister came during the second day of spring shearing. Frank, the assistant shearer, came in and announced, "You have an addition to your herd." I went out to see, then set about moving a kidding pen to the shearing room, which is normally the maternity area and is warmer. By the time I returned to get the mother and kid, there was a twin. The mom, Jane Austen, is a large, somewhat cantankerous doe who has had one kid before, two years ago. From the first she did not seem as interested in the female as in the little male. She licked them both, but in a desultory way, without the focus and almost frantic attention I find reassuring. I hovered around, attending to the little chores of birthing. I held each kid up to a teat and put a drop of milk on its mouth, then let the mom finish drying the kid. Later I held the female kid up to drink again. Jane seemed distracted by the noise of the shearing machine in the center of the room and kept looking around at it, but it was unquestionably the best place for her pen. I left to prepare lunch for the shearers, and when they came into the kitchen they reported they had held each baby up to the doe and each had taken some milk.

During lunch visitors came to see the goats, friends of friends in another part of the state. I suggested that tomorrow would be better after the shearing, but they pleaded. They were to leave the island the next day and so much wanted to see the goats. Their little girl, particularly, would be so disappointed not to see them. I left the table and led them on a tour of the barn. When we reached Jane's pen, the little female's mouth was cold. I strung up a heating lamp and ran back to the house for a syringe and glass, then milked some colostrum from Jane. The baby wouldn't swallow it, so I returned to the house for dextrose and more syringes and needles, and gave the kid several injections of the sugar solution. I ignored my visitors and hoped they would find that simply watching was sufficiently interesting. The dextrose helped but the tiny kid still refused to drink, which sent me back to house again for the stomach feeding apparatus. I carried it in a pan of hot water to keep it warm in the chill air. Frank helped me by holding the infant, and I was grateful. Tube feeding is vastly easier if done by two people, and we succeeded in getting a couple of ounces of milk into the baby. At that point the visitors decided they were in the way and left, not unsatisfied, I think, after their tour and a bit of drama.

While taking a break from shearing, Larry, the head shearer, climbed into the pen and held the baby up to the doe's teats, keeping a finger in her mouth to stimulate a sucking response. Again she refused to suck. We finally let her rest; after all, she did have two ounces of milk in her stomach. Later, when the shearers were at the Karen and Lee's, shearing their sheep, I started dinner and then checked the barn, certain I'd have to do another tube feeding. But the little one raised her head; it was the first real sign of liveliness that I'd noticed. I held her up to a teat and she drank, and I soon returned to the house, hoping the crisis was over.

Now I've had several days of alert with little happening. I will these unborn kids to stay in the womb as long as possible; a too-early arrival greatly reduces chances of survival. Do I expend energy needlessly on my hoping, my willing the births to come out right? Life continues as it will regardless of my mental meddling, and perhaps one day I'll accept it as it comes and be the more serene for it.

The twins are out of danger; I no longer poke a finger in their mouths at every turn to assure myself they're getting enough nourishment. Still, they're too frail to remove from the kidding pen, where their mom is increasingly restless. Since it's not frigid this year at kidding, I can leave for an hour's walk with the dogs. However, I accept no invitations at this time; I cancel any appointment that may take longer than an hour and even the ones I keep are with the provision that a sudden birth may keep me away. The one-way intercom, connected with the barn, buzzes beside my bed while I sleep. I'm never far from it except when I go to the post office or grocery store twice a week or walk the dogs.

I write this with my dogs at my feet; Lily rests against my outstretched legs and Molly lies near my chair. I've moved the intercom into the room where I write. When I listen to it and hear the sounds of the goats milling around in their pen, grunting at each other, my mental image of them is so vivid that I could be watching a movie on my internal screen. My mind's eye sees them settled in the hay, some standing, some stretched out on their sides. I see them drink from the canning kettles when I hear the clink of metal as the goats bump against them. I hear them snort and cough and I see it all. When a doe bellows in labor, I see her before I reach the barn. In the very early morning when it's still dark, I hear only occasional rhythmic grunting. At seven a few barn sparrows begin to chirp, and the chirping gathers in volume and momentum until it's a wakeup chorus. This fascinates the cat, who looks at the intercom, puzzled but twitching the end of his tail as if contemplating a pounce. It's a happy way to start the day.

A few days ago the weather turned briefly mild. Listening to the intercom, I realized I heard nothing. I went out to check the goats and discovered they had all wandered down to the far end of the pasture near the rock wall at the southernmost boundary, some three hundred and fifty yards from the barn. This is a worry because it is entirely possible for a doe at the point of giving birth to follow the herd to the far reaches of the pasture, have her baby on the wet, thawing ground and return with the rest of the does when they amble back to the barn. The baby would be left behind. These days when the weather is fair and I'm not working in the barn where I can keep an eye on the herd, I shut the does in.

Epilogue

YESTERDAY the wind shifted from the southwest; it passed West Harbor, rounded Denny's Bluff and Boyers Bluff, then chilled Schoolhouse Beach as it made its way toward Rock Island until finally it blew from the northeast. I watched its effect on the water from my vantage point at the house I rent from Jim during the summer, high above Lake Michigan on the north side of the island. The changing winds riffled the surface, which shattered into pieces of sparkling mosaic, rippling in place as if not knowing which direction to take. Today slow gray-green waves surge up and roll to shore; they seem to bring autumn.

The late spring birthing turned out to be our last, with the exception of a few unplanned births the next year. Peter and I never made the trip to Canada to inspect the commercial mill there, and he could never discuss the future of the goat project because of his other business concerns, which he never specified. Realization came gradually that our business would never develop. Over the months I accepted Peter's refusal to discuss our goat project; there was no use in persisting when he was not ready. I felt sad at this because it seemed increasingly that he had lost interest in raising goats and ultimately creating a business based on their fiber. In those days I often felt without direction, without a goal. But I found the animals so fascinating and such a joy to be among that not for a moment did I ever regret what I was doing. Financially it made no sense at all for me to invest productive years in something that paid virtually nothing and would end with no material return. But I believe that richness of experience in the end far outweighs other forms of income, at least for me, and financial uncertainty is a small price to pay for the full life I've been living on the island. No matter how discouraged I felt at the gradual unraveling of my dream, I understood that the sojourn with the goats was a magical experience that I could not have easily provided for myself. I was always aware, too, that it was my choice to stay on the island and care for the goats. No one was making me do it; Peter could have hired someone to maintain the herd.

Increasingly I felt that I was waiting for a signal from Peter that would tell me what I knew was inevitable: that my idyll on the farm had to end. Finally Peter announced that he was negotiating the sale of his manufacturing company and was planning to retire, which would leave him free to move to the island. He and Jeannine, in fact, had been discussing various projects and farming ventures—including a mohair-processing business—that the two of them could develop together once they were on the island.

Peter and his family did move to the island, but I could not leave then because the people who were living in my condominium had a year left on their lease. I was relieved at this, because I loved being there and dreaded searching for a job—and ultimately being confined to an office—in Santa Fe. I moved into the guest apartment at the farm except in the summer months, when it was needed, and continued to work with the goats. During that period my brother's great capacity as a restless innovator and visionary drew his attention to New Zealand. Fortunately New Zealand also intrigued my sister-in-law. They and their children visited and they immediately applied for residency permits. While waiting for the permits, Peter launched into a remodeling project that enlarged the 1,600-square-foot farmhouse to 4,000 square feet. The permits were ultimately granted, and shortly afterward Peter and Jeannine left the girls in my care while they traveled to the South Island of New Zealand to look for farmland and the site of their new life and latest adventure. As I write this, Peter and Jeannine are planning to move in a few months' time. They will close the island property and put it up for sale six and a half years after we first acquired the goats.

As I walk through the woods, the stained-glass green and yellow of the canopy of leaves filtering the sunlight, I'm moving through a vast yet intimate cathedral. In three months' time all traces of this brief summer will have vanished; leafless birches will again pierce the blue sky, their luminous white trunks casting long shadows on the shimmering snow underfoot. But by that time I will no longer be on the island, which seems unimaginable, for I do not know how I can extricate myself. In the end, I've burrowed into the island's fabric of nature and warm-hearted community more deeply than I ever thought possible. My main concern now is to find homes for Ariadne and Desdemona, for fat, agreeable Musetta and Carmen and for José and the rest of the goats. I wander into the pasture on frequent visits to the farm and am eager to move back and, in the short time that remains, to spend more time with the animals, to capture them indelibly in my mind and to take with me some of their serenity and goodwill.

About the Author

Susan Basquin spent her childhood in Evanston, Illinois, vacationing in the summers in Door County, Wisconsin. She majored in art history at Smith College, and spent her junior year in Spain. After college she lived and worked in New York City, primarily in public relations and for magazines. After ten years, she moved to New Mexico, first to Albuquerque, and later to Santa Fe, where she worked at a weekly newspaper, most of the time as a writer and editor. She now lives in Santa Fe with her two dogs and three cats. She works as a technical editor and writer and on her own writing projects and volunteers at the Santa Fe animal Shelter.